Internet, Society and Culture

Internet, Society and Culture

Communicative Practices Before and After the Internet

Tim Jordan

B L O O M S B U R Y
NEW YORK • LONDON • NEW DELHI • SYDNEY

Bloomsbury Academic
An imprint of Bloomsbury Publishing Plc

175 Fifth Avenue
New York
NY 10010
USA

50 Bedford Square
London
WC1B 3DP
UK

www.bloomsbury.com

First published 2013

Library of Congress Cataloging-in-Publication Data
Jordan, Tim, 1959-
Internet, society and culture : communicative practices before and after the Internet / by Tim Jordan.
p. cm.
ISBN 978-1-4411-3487-5 (hardback : alk. paper) 1. Internet–Social aspects.
2. Communication–Data processing. 3. Information society. I. Title.
HM851.J68 2013
302.23'1–dc23
2012030070

ISBN: HB: 978-1-4411-3487-5

Typeset by Newgen Imaging Systems Pvt Ltd, Chennai, India
Printed and bound in the United States of America

Contents

Acknowledgements

I'd like to thank the following for making whatever their intended or unintended contribution was. Of course, the book (including all its faults) is my responsibility.

Katie Gallof has been a very helpful and interested editor. Two anonymous referees helped ensure I made clear the methodological choices underpinning this project.

Open University colleagues helped by listening as I started to sort out these ideas as the project began. In particular Richard Collins, who showed me Esther Milne's book *Letters, Postcards, Emails* (which nearly stopped me in my tracks when I thought it was the project I was embarking on), and Steve Pile who both helped intellectually and also in other important ways when my personal life was difficult. Jenny Robinson offered a place to stay when I was in-between houses which was both generous and deeply needed. Much of the writing up was done while on a sabbatical from the Open University which helped significantly to recover from being Head of Department. The Sociology Department at the Open University also supported this research with a small grant that enabled a pilot project into the research on letters.

Many thanks to Donna Haraway and the History of Consciousness Department at University of California at Santa Cruz who offered a very congenial and challenging home, and I apologize to them for not giving enough back. Between the trees of UCSC's campus and trips to watch Steamer Land and Mavericks, this was an environment that helped me to think-with.

The British Academy supported this research with a small grant without which the archival analysis of letters in Australia would not have been possible. The librarians in the State Library of Victoria were magnificent. The Mitchell Library in Sydney was also helpful. My friends and family in Australia offered a counterpoint to the long days immersed in the nineteenth century with twenty-first-century wine, food and company.

The research group Cultural Production in the Digital Age run by Tarleton Gillespie and Hector Postigo provided an important intellectual environment and the chance to talk through ideas with other better informed colleagues.

New colleagues at King's College London in Culture, Media and Creative Industries and in Digital Humanities have provided a stimulating environment. Students on the Masters in Digital Culture and Society at King's College London have also provided many digital natives who've corrected and expanded my knowledge.

My friends Nick, Rahman, Liam, Rodolphe and latecomer Joth have proven how virtual friends can become some of the best of friends. And to my guild AS which provides much amusement even when Anders tricked me into becoming guild leader again.

Finally, this goes to Matilda and Joanna with love.

1

Before and After the Internet

Introduction

The nature of culture and society changed at the end of the twentieth century, as novel forms of communication dependent on internet technologies came into widespread use. With the internet came not just email, electronic discussion boards, social networking, the world wide web and online gaming but across these, and other similar socio-technical artefacts, also came different identities, bodies and types of messages that changed the nature of communication and culture. The following arguments explore interrelations between the rise of the internet and different identities, bodies and messages in communication and examine their effects on twenty-first-century cultures and societies. The focus is on the practices that make the sending and receiving of messages possible and how these practices have changed. This will be done by comparing a case study of pre-internet communication using early nineteenth-century letters with a case study of deeply immersive internet communication using online virtual world gaming. This will lead to consideration of the meaning of changes in communication brought by internet technologies for wider cultural and social change, particularly in the normalization of communicative anxiety.

Such a project explores the nature of communication after the rise to mass use of internet technologies. In this sense, being 'after the internet' is not the same as being without the internet but instead refers to how communication operates once internet technologies are integrated into it. The first step of this project is to consider the claim that there has been social and cultural change related to internet technologies, and, to do this, it is useful to look at a puzzle about metaphors and analogies between the non-virtual world and virtual phenomena. Such metaphors are nearly always based on a familiar phenomenon from the non-virtual world (e.g. letters) that is then applied to an aspect of the virtual world (e.g. email) to explain or introduce the latter. The puzzle is that such metaphors often appear obviously and intuitively clear, allowing what seemed novel and puzzling to be understood as familiar and obvious, yet after some consideration such metaphors usually turn out to be significantly misleading. What at first seems to be an insight turns into a failed interpretation, and in doing so offers an indication of cultural and social changes that have come with mass use of the internet. To see this, we can look at two examples of the difficulty of comparing what seem, at first glance, to be the same acts conducted in the online and offline worlds; burglary and street protest.

Metaphors and their failures: the metaphor of burglary

Hacking, or to some cracking, refers to the act of breaking into someone else's computer remotely.[1] As has often been said in computer security circles, the only way to be certain someone cannot access your computer illicitly is to lock it in a secure room that has no access to the internet and then allow no one else access to the room. Once a computer is connected in some way that allows other computers to access it then the chance that someone can break into that computer is always there, whether it is secured by password or firewall or these and all manner of other computer locks.

The history of cracking offers many varied and increasingly complex ways of breaking into a computer (Goldstein, 2008). These range from the cracker who gained access to the Duke of Edinburgh's email account by guessing that the password would be 1234, to the production of a program that automated breaking into accounts on Microsoft networks, to the 'playful' types who break into websites and rename them (the Central Intelligence Agency renamed to the Central Stupidity Agency, for example) (Taylor, 1999, p. 72; Jordan and Taylor, 2004, pp. 111–14). From the beginning of cracking, it seemed that an obvious metaphor for it was burglary. The success and failure of this metaphor can be examined drawing on a formative time in the history of hacking and the internet; the 1990s period of cracking, which is now sometimes called its 'golden age,' before the more criminal and geopolitical phase of the early twenty-first century (Sterling, 1994; Menn, 2010; Poulsen, 2011).

Burglary makes sense as a metaphor for cracking because it captures its key characteristics. There is the sense of the illicit and the need to break something to pass a boundary, and it further captures the sense of an invasion of space by someone not meant to be there who is gaining some advantage. Most important of all, burglary is well known and easily understood and can thus make the strange into something commonplace. The burglary metaphor was popularized primarily by computer security professionals, who were often desperate to explain to the internet-illiterate in the 1990s what cracking meant. It was therefore an advantage that burglary not only explained what cracking meant but also rang the alarm bells that many felt were necessary. For all these reasons, burglary became during the 1990s, and remains, an oft-used metaphor for cracking. However, when more closely examined, the metaphor begins to appear confusing and, under sustained analysis, misleading. Examining this makes clear that the burglary metaphor for cracking does not work as an accurate representation of cracking but instead its inaccuracy functions to establish moral judgements about cracking.

What criminal breaks into someone's home and steals their television by taking an exact copy of that television, leaving both the victim and the criminal with a television? No burglar does this of course, it is impossible, but this is what a cracker does. Crackers do not, usually, remove digital objects, they copy them. This point has often been noted when criticizing burglary as a metaphor, and it is the first step in seeing that digital burglary is not really like physical burglary. There are some other differences that quickly appear as more is found out about cracking. For example, crackers hold publicly advertised conferences, which is not a usual practice for burglars or criminals, and crackers sometimes ring up the sites they have cracked to advise systems administrators on their failures and how to fix the problem, which is again not a usual or familiar criminal practice. The latter is a practice still alive in 2011 when the hacking group LulzSec broke into part of Nintendo but reported the breach to Nintendo because of their proclaimed love for Nintendo gaming (Winterhalter, 2011). Both in the types of actions taken – copying versus taking – and in their attitudes to what is done – open discussion versus secrecy – crackers do not seem to fit an obvious understanding of what a burglar does.

The recognition of such inaccuracies sometimes leads to the reformulation or extension of the metaphor of burglary to try and make it stick. A computer-systems manager Bernie Cosell offered the following adjustment:

> There is a great difference between trespassing on my property and breaking into my computer. A better analogy might be finding a trespasser in your high-rise office building at 3am and learning that his back-pack contained some tools, some wire, a timer and a couple of detonation caps. He could claim that he wasn't planting a bomb, but how can you be sure? (Cosell, cited in Jordan and Taylor, 1998, p. 772)

We can note that this retains many elements of burglary, breaking and entering, particularly, but shifts the sense of what occurs after breaking in. A then-UK-government official, Mike Jones, attempted a similar adjustment both trying to retain the sense of threat and danger involved in burglary but acknowledging that burglary and cracking are dissimilar.

> Say you came out to your car and your bonnet was slightly up and you looked under the bonnet and somebody was tampering with the leads or there looked like there were marks on the brake-pipe. Would you just put the bonnet down and say 'oh, they've probably done no harm' and drive off, or would you suspect that they've done something wrong and they've sawn through a brake pipe (Jones, cited in Taylor, 1999, p. 111)

Warming to his adjustment of the metaphor, Jones enunciated a second reinterpretation of cracking as burglary shifting it further to forms of illicit access that carry an implied threat, this time moving from cars to airplanes.

> Say a maintenance crew arrived at a hanger one morning and found that somebody had broken in and there were screw-driver marks on the outside casing of one of the engines, now would they look in side and say 'nothing really wrong here' or would they say, 'hey, we've got to take this engine apart or at least look at it so closely the we can verify that whatever has been done hasn't harmed the engine.' (Jones, cited in Taylor, 1999, pp. 111–12)

From a house to an office to a car and then a plane, the metaphorical position of the computer that is being cracked shifts as each attempt tries to retain an ethical sense of what a 'crack' means while failing to equate the physical and digital realms.

The difficulty with these metaphors points towards two conclusions. First, in this case the metaphors are primarily a means of establishing an ethical view of cracking, not of representing cracking accurately (Jordan and Taylor, 1998, pp. 770–5). Second, and this is the key present point, a metaphor that seems obviously and intuitively correct between acts in physical space and acts in digital space does not work and is significantly misleading. Things appear to be different when a seemingly similar action is taken over the internet and in a house. However intuitively similar these acts are, they are in fact quite distinct. A second example will help further explore this point in the creation by hacktivists of mass civil disobedience on the internet using the model of the street protest.

Metaphors and their failures: the metaphor of protest

In the mid-1990s, a number of political activists began to explore the consequences for civil disobedience of the emergence of the internet. Activists had not only adopted email, electronic fora and other communication possibilities produced by the internet to help organize but had also begun to think about how to take direct action online. Activists began to explore and develop ways in which familiar offline protests such as boycotts, blockades and other forms of non-violent direct action could be recreated in online environments, leading to a politics called 'hacktivism.' Out of this came one particular strand of online direct action in an attempt to recreate mass street demonstrations online. The logic was that if information flows have become as important to centres of power as physical flows, then there needed to be a way of blocking information flows that was equivalent to the ways street demonstrations obstructed physical flows (CAE 1996, 10–15; Jordan and Taylor 2004, 67–74).

The equivalence between mass street and mass online demonstrations can most clearly be seen in examples where a virtual demonstration is timed to coincide with street demonstrations. The anti-World Trade Organization demonstrations in Seattle in 1999 were both virtual and non-virtual. The actions that made most of the headlines were those in the streets, where demonstrators sought to block roads with the mass of many bodies so that conference delegates were unable to get into the conference building. For example, early in the morning of 30 November, demonstrators took over the main intersections around the conference venue, several different marches then brought more demonstrators into the area. With numbers of demonstrators much higher than expected, they had effectively blocked all the streets and the key intersections around the conference venue, blocking some police inside a ring of protesters. This meant that conference delegates could not pass through the demonstrators. Police attempted to break the encircling demonstration and join up with police already trapped inside. Beginning as a non-violent demonstration, its very success led to police attempts to forcibly break the lines to remove demonstrators. The presence of violent protesters also contributed to the emergence of the now-iconic pictures of destruction and fighting that led to the demonstration being called 'The Battle in Seattle' (Gautney, 2010).

Here is a classic street demonstration which effected its politics using human bodies, and ingenuity, to physically block space and to prevent others using those spaces. At the same time, human ingenuity was being used to block the electronic wires supporting the WTO conference. A protest group called the 'Electrohippies' (also Ehippies) set up an action to run concurrently with the street demonstrations that would allow anyone unable to physically be in Seattle's streets to attend virtually. The Electrohippies set up a means of bombarding the WTO computer network with requests; essentially anyone could participate by going to a website set up by the Electrohippies and by clicking on a link that then automatically repeated requests for certain pages from the WTO site. The Electrohippies claim that this was a successful action, believing that they stopped the WTO servers on 30 November and had 450,000 uses of their links (Jordan and Taylor 2004, pp. 74–9). And such actions continue having now entered the repertoire of political activists. For example, in March 2010, the Electronic Disturbance Theatre organized a virtual sit-in at the President of the University of California's online portal to coincide with street demonstrations against fee increases and other issues at the University of California (Goodin 2010).

Yet the validity of such political actions was challenged by other online activists who did not accept the equation of an online blockade with an offline blockade. These activists rejected the metaphor and in so doing exposed its confusions. The fundamental criticism was that the body that helps constitute a blockage in a street is not the same as the 'body' that blocks connections on the internet. One way of seeing this is to note that it is easier for one person to block connections to a particular site on the internet than it is for thousands of people to do so. Attacks on websites that flood them with data and so block their connection to the internet making them disappear are well known as denial of service (dos) attacks and have rendered invisible many major online presences. Most of these attacks are conducted by one person or a few automating the production of information requests to the target, for example, by infecting a wide range of computers with 'zombies' that can be set to suddenly produce large flows of requests to connect to the one target at the one time. All this can be done by a single person. Such attacks are characteristic of the dos attacks in 2010 orchestrated by activist group Anonymous to strike back at organizations they felt were attacking WikiLeaks. Though several Anonymous members participated – for example, five were arrested in the United Kingdom in late 2010 for their alleged participation – each attack utilized a massive reproduction of information, reputedly based on the software package LOIC, and thus multiplied information hugely over the number of bodies participating. In this way, such sites as MasterCard were claimed to have been slowed and taken down for short periods (Addely, 2010).

But the actions of those like the Ehippies or Electronic Disturbance Theater (EDT) must avoid the accusation that they are single or just a few people, because they need the mass of bodies to be able to claim to be a public protest that expresses a legitimate political claim precisely because it is, like a street protest, a mass. Somewhere embedded in the idea of mass street protest is the legitimacy conferred on this protest by the numbers of people involved, and this political claim needs to be translated into mass online protests. Such online protests therefore often utilize technologies that limit the powers of the internet and avoid the ability that Anonymous and others using dos attacks have in multiplying protesting bodies rapidly and massively. The virtual protest body only corresponds to a street protest body by limiting the capabilities the internet offers.

This leads to the paradoxical situation in which the most technologically advanced mass protests must utilize impaired forms of technology to retain political legitimacy. The Ehippies protest offered two links to click on depending whether a protester had a fast or slow connection; the fast link reloaded six pages on the WTO site automatically while the slow loaded two. The Ehippies could have set these reloads to be much higher or could have launched automated dos attacks but they had to utilize technology that was less effective at taking down the WTO site so that they retained the political legitimacy conferred by being a *mass* protest.

The issue of political legitimacy and its different manifestations online and offline demonstrates how misleading the equation of street and online mass protests is despite that equation having been a basis for the creation of this political tactic. There are also other differences that can be quickly found. For example, online protests have been criticized for their ease and lack of danger, the comparison between clicking a mouse and running from riot police suggests a very different level of commitment between the two protest types. A further difference is that one of the key issues with a mass street protest is the logistics required to get a large number of people together in the right place at the right time; again this differs radically with online protests where such logistics generally involve turning on a computer, perhaps after having received notification through an automated email list (Jordan and Taylor, 2004, p. 80). The more the nature of a mass online protest is probed, the more such protests seem different to offline mass protests. The use of classic civil disobedience to present online protest as a metaphor, such as when the Critical Arts Ensemble calls for 'electronic civil disobedience,' turns out to be initially attractive but substantively misleading.

Despite the seeming immediacy and accuracy of many metaphors of offline for online phenomena, they turn out to be misleading in case after case. We can enter 'chat' rooms and 'talk' to people when in fact we are typing, and everyone can talk all at once and retain full communication. We 'go' to 'places' without moving an inch from our chairs, with just the pixels rearranging themselves on our computer screens. And it is not that we fail to move on the internet, just that using the space we know that is not involved with internet technologies as a metaphor or guide to the spaces that internet technologies are part of producing, will initially beguile us but also mislead us. This failure is a clue to the larger issue; to what extent are social and cultural norms, ethics and practices different or similar when they are or are not dependent on internet technologies?

Moreover, such metaphors contribute to the way debates over the effects of the internet have often become polarized. The misleading understanding analogies and metaphors offer us, allow the comforting claim that something that seems new is actually familiar. When they fail, as they habitually do, they also then open the door to the opposite claim that something entirely new that supersedes the old has appeared. Analysis can then be caught in an opposition between claiming nothing 'really' new has appeared or its opposite that something radically new has appeared, rather than comparing and delineating what is new and what is the same. The present argument pursues the change that the failure of metaphors like street protest for online protest suggest exist, but does so to be clear that there are likely to be both similarities and differences between communication before and after the advent of internet technologies. To explore this, there will be a need to grapple with both the mess and complexity of variable interactions between technologies, groups, individuals, signs, actions and more through which communication is lived and routine and habitual practices that in their repetition also make up communicative practices. The first stage in this analysis is to take this clue that failed metaphors offers and turn it into a hypothesis based on existing studies of communication and the internet.

A hypothesis of communicative practice after the internet

The failure of metaphors drawn between non-virtual and virtual spaces suggests differences between the two. To develop this difference and focus it on communication I will propose the following hypothesis. The difference between online and offline can be taken as a sign of the existence of two concurrent and interacting communicative practices: one communicative practice was developed in Western societies prior to the emergence of internet technologies and is familiar to the point of being nearly entirely taken for granted; and, there is a second communicative practice that has arisen with the emergence of internet technologies and has been rapidly developed and assimilated by many.

A hypothesis here can be thought of as the first suggestion that explains some of the behaviour indicated by the failure of metaphors between online and offline and that is also being studied in the quickly grown field of internet studies. Further, 'hypothesis' can be understood in its older meaning from Ancient Greece when a hypothesis was a summary of the plot of an ancient drama. Such hypotheses acted as something like a preface giving the story, its setting, main characters and the context for a play's initial production (Vlastos, 1994; Kovacs, 2005, pp. 384–5). If the modern drama at stake in this book is 'Has communication and culture changed with the arrival of the internet?,' then another way of understanding what I mean by hypothesis is that it summarizes the story's plot by outlining its main meanings and conceptual protagonists. A hypothesis of this type is well served by telling the story once in a small way and then retelling it with conceptual and empirical complexity.

The hypothesis, or storytelling, has implicitly begun in the identification of the possible difference between phenomena that are offline and online. The phenomena that were used to suggest this possibility were each based centrally on communicative contexts, whether that of illicit communication to computers or of destroying communication to targeted protest sites. Putting communication and practice together suggests exploring sets of social and cultural relations played out through material practices that establish ways in which people may communicate. Material practices here refer to repeated actions emphasizing that they involve empirical and tangible resources. Communicative practices set out how we, normally unthinkingly, stabilize the elements of communication and how in daily practice we easily answer questions such as: What is the identity of the sender? How does the receiver know this message is from the sender? What is the identity of the receiver? How is the self-identity of the message maintained? How can the message be read? Communicative practices focus not on the moment of transmission, when meaning passes from a sender to a receiver, but instead on the means by which such transmissions are created and are able to be repeated across masses of people. The hypothesis being developed can address communication before and after the internet by setting out the nature of communicative practice prior to internet technologies and then comparing them to communicative practice dependent on internet technologies. I will briefly sketch in these stories.

Communicative practices prior to the internet began from those derived from face-to-face communication. These are based on a collapse between a physical body and an identity that then legitimizes authors, recipients and messages. All these elements can be swiftly negotiated when face-to-face because all parties to the communication are present. The development with face-to-face communication is one in which there are masses of authors and recipients who are at a distance to each other. This creates the problem of authorizing a communication when the author and recipient's identities and physical bodies cannot be automatically known to be the same. The era of communicative practices prior to the internet derives from this development from face-to-face forms of authorization of communication because of the separation of communicative identities from communicating bodies that is required by at-a-distance communication.

Communicative practices for at-a-distance communication create transmission based on authorizing the coincidence of body and identity using material practices involving letters, stamps, signatures, voice and so on. Where being at a distance pulls the body and identity of face-to-face apart, these practices of communication now authorize a reconnection of communicative identity to body by creating a communicating body. One example of this is the signature, which is a practice in which the performance of inscribing certain shapes produces a coincidence of name and body, because the signature establishes that the body of whoever is named in the signature touched and produced a message, much as if it had been spoken aloud. Signatures can be traced back to royal practices of seals in which the person of the monarch, the body of the king, was legitimized as the author of a communication (whether the communication was personal, state, legal or other was often reflected in a different seal and signed name). The physical body of the monarch was attached to a communication through a device which could not be activated unless that physical body was tangibly at the communication (hence, also the practice of seals, particularly personal seals, being made into a ring and so being intimately attached to a body that 'owned' that seal) (Clanchy, 1993, pp. 51–77). The development of written languages and the extension of literacy integrate into a range of different media this way of stabilizing the transmission of messages.

The central dynamic of pre-internet at-a-distance communicative practice is the use of written language on a variety of media that is legitimated as coming from one identity through practices and technologies that authorize the message as having come from a physical body co-terminus with the sender's identity. The problem of separation of identity and body is solved through a range of practices that stabilize the two as one and legitimize certain performed marks as mediators which state that a certain body is attached to a certain identity that produced the message. For example, characteristic styles of handwriting, or particular styles of language use, or certain types of salutations and ways of signing off can all be interpreted, when they are known and repeatedly used, as stabilizing a body and identity as the author of a message, just as seals and signatures can. Through these practices the physical body and identity of communicants is not reflected but is produced through fragments that stand in for and are taken to authorize the communicant.

Such a complex set of attachments stabilizes and underpins communication through the pre-internet era. This form of communicative practice is flexible enough to continue through a range of new developments. For example, the telephone created little disturbance to the basic structure of communication because it provided another means of authorizing a physical body through fragments found in the timbre, intonation and characteristics of someone's voice. Other forms of communication, such as the telegram, posed more challenging problems with its restricted vocabulary undermining strategies of identifying through performances within messages and carrying only the barest of signatures. While telegrams are part of pre-internet communicative practice, this may only be because of the exceptional nature of telegraphic messages within the vast weight of other such forms of communication.

In contrast, internet-dependent communicative practices no longer stabilize communication through material performances that merge identity and body. Instead, the body as a means of stabilizing identities fades away as a locus of authority because the internet characteristically offers faulty and suspicious markers of identity leaving the body overshadowed by practices in which styles of messaging stabilize a communication. This is caused by a communicative context in which markers of identity – such as email address, forum name, Facebook name – are unstable and themselves need to be authorized. The extent of the instability of this communicative practice may be a marker of its novelty, perhaps first initiated in telegrams but only coming into widespread use towards the end of the twentieth century (Standage, 2007). In this communicative practice, styles of message become evermore important and so the symbolic content of messages becomes more closely integrated into the means of stabilizing communication. The invention and use of new forms of text communication, such as emoticons or 'text speak,' alongside choices about what new styles one does and does not use become evermore important to authorizing who has created and who is receiving a message and whether that message can be understood.

For example, in many online fora if a post is made under one name and is followed immediately by another post using a different name whose content either strongly agrees or disagrees with the first post, many will immediately suspect that the two posts come from the same person. The suspicion is that 'trolling' is going on in a deliberate attempt to spark an argument for the sake of the argument. In such a case, the identity-markers claim two different communicants are at work but the style of the interaction is read contrary to the identity-markers to instead assume that there is only one person sending messages. Other markers may buttress this kind of reading. For example, if a forum offers a post-count of the number of times a particular name has posted a message then if the first post has a high post-count and the second with a different name has a very low post-count (or vice versa), this will emphasize suspicion that the two are from the same identity (one from the user's normal and highly used account and another from a rarely used one) (Donath, 1998). A second example is the experience of someone who sends mass advertising emails, or spam, illicitly taking over an email address and routing the advertising mails through that address to the person's contact list. In this case, what may well be a very familiar email address associated automatically with a particular body offers up a style of message radically at odds with the expected style of the communicant. If someone you know and whose email is familiar suddenly sends advertisements for viagra, for enhancements to body parts or suggests new dating sites, then most will immediately assess the email from the style it offers rather than from the email address. Styles of communication, particular themes, forms of writing, characteristic examples and other such markers of a particular way of communicating, have become the means of stabilizing communication from a sender to recipient.

Two communicative practices, each made up of many and varied material practices, have been hypothesized and their differences revolve around the ways in which messages are legitimized as coming from one identity and are received by another identity and ways in which the nature of the message can be constructed and maintained. The key shift is from the body to style. But these practices should not be thought of as mutually exclusive or sequential in the sense of one inevitably taking over from the other. If we take the viewpoint of a particular individual then it is not that any one person will experience these two communicative practices separately but s/he is likely to experience the mess of everyday life flitting in-between and within two sets of communicative practices that themselves pull in different directions, have different standards and are potentially dissonant to one another.

This hypothesis is a first approximation to understanding what the clue found from the failure of metaphors between offline and online means, all it does is to open a way forward towards understanding what has happened to communication. The hypothesis tells the first story. The question now becomes how best to study this hypothesis? If this first telling of the story leads to the suggestion of two different cultures that sustain different ways of transmitting and receiving messages, then the task becomes finding a way both to understand and test the story. It becomes a project of retelling the story in a more complex and empirically grounded way to see whether the story remains coherent and sensible or whether it comes apart and disintegrates. This however sets a number of interesting methodological problems.

Methods: theory, comparison, difference, intensity

A theoretically complex and empirically informed retelling of the hypothesis is necessary. What methods can do in this context is to give a specific form to the retelling that offers the greatest chance of finding out whether the story remains coherent once it becomes complex. A number of components to such a method are needed. Theoretical complexity has to produce an understanding of communication that stretches from pre-internet to internet-dependent historical periods. While the latter is not so long, the former stretches far back into communication history. Analysing both periods for their communicative form then provides a basis on which comparison may be conducted. However, confusion is possible because internet-dependent communication does not replace and is not mutually exclusive from pre-internet communication, but rather continues to exist alongside and to interpenetrate with internet-dependent practices. One way to minimize this potential is to take a case study of pre-internet and one of internet-dependent communication that are as clearly representative of each as possible. To ensure comparison can be done meaningfully, it is important for the case studies to acutely express the specific nature of their communicative contexts. It will be useful to clarify these two aspects of the method being employed; comparison and difference.

To create the possibility of comparing pre-internet and internet-dependent practices, a theory of communicative practices that applies to both will need to be articulated. The hypothesis already implies significant conceptual work because of the extensive use in it of terms that have themselves been subject to considerable conceptual controversy over, at least, the last 30 years. What is meant by message? What is meant by identity? What is meant by body? What is meant by transmission? Is a theory of performativity (and which one) needed or is a theory of practice or both? These kinds of concepts need to be taken into account to fully understand what the hypothesis might be claiming, and this theory also needs to outline a view of communication that can confidently be used across the eras of pre-internet and internet-dependent communication.

This theoretical task will be undertaken by drawing on two related but not always connected areas of analysis. Communication studies will be a key resource, particularly in the debates around the meaning of transmission of messages and cultures of transmission that have developed around the term 'communicative practice'. The work of Carey, Peters and others will be taken up here. Not always connected to this work have been developments in cultural studies around language and the body which also deal with issues of how communication is created and maintained. Connecting this cultural studies work, particularly as derived from Derrida and Butler, to communication studies will create a complex view of communication that takes many issues back to an abstract foundation in the ways in which the transmission of messages is created and maintained (see Chapter 2). While the meaning of communicative practice will be defined abstractly, the theorization also requires that each communicative form will be located in particular times and spaces. Such a theory outlines how it is possible that the particular communicative practices that operate in pre-internet and internet-dependent times and spaces can be defined and compared. Abstractly both pre-internet and internet-dependent communication are communicative practices, but in practice they exist in specific times and spaces and have particular forms.

To counter the potential confusion produced by these factors, the empirical evidence will be collected from case studies of intense forms of communication. Intense, or 'extreme,' here does not mean unusual but rather communication that is heightened by the particular time and space it exists in. Comparing two case studies of intense forms of communication should provide the clearest possible examples of communicative practices and allow conclusions about their differences and similarities to be drawn. For similar reasons, a historical view that starts prior to the internet and traces the emergence of internet-dependent communication is potentially subject to confusion given the more gradual change over time provided by a historical narrative rather than the abrupt comparison of communication from two different eras. This does not mean a historical view of communication is either inaccurate or untenable, only that the more direct route to understanding the differences of communication at stake here is to focus on two clear comparable case studies. The final element of the methods underpinning this book is then to define the case studies and why they are 'intense' forms that offer clear evidence of communication in their era.

For pre-internet communication, it seems advisable to move prior to the telegraph to remove the possibility that the telegraph first brought internet-like communication (Standage, 2007). One way of identifying an intensive form of pre-internet and pre-telegraph communication, that is also amenable to research, is to look at communication in letters that is radically at a distance. If the letter writers and receivers are themselves so distant that effectively all communication between them is carried by letters, then examining those letters will put the researcher in as close contact as is possible to pre-internet communication by offering the researcher nearly all that constituted the communication when it first happened. One case study that fits this specification is letters to colonial Australia in the early nineteenth century, and archives of such letters were examined (see Chapter 3). In this case, not only were several series of letters between individuals found but also case studies of business letters and of state letters. These were supported by work in other archives to form a case study of pre-internet communication. The first case study is then an archival study of letters to and from colonial-era Australia.

The second case study needed is of 'intense' internet-dependent communication. While it might seem obvious to compare letters to email, and clearly this can be a productive research method (Milne 2010), there is also the difficulty of further confusion caused by similarities in form between emails and letters. In addition, more immersive environments are possible based on internet technologies, environments which cannot be imagined without such technologies, whereas email lies at least partly within the imagination of letters. A communicative environment unimaginable before and without internet technologies would be an intensive form of internet-dependent communication, and these can be found in virtual worlds. Such worlds offer physically dispersed individuals access to an online space that presents itself as a new world, some of the best known of which are World of Warcraft (WoW) and Second Life. The second case study takes up this sense of intensity by conducting ethnographies in virtual worlds, primarily in the massive multiplayer online games WoW and Dark Age of Camelot (DAOC). These ethnographies aim to produce empirical evidence for the ways communicative practices that are dependent on internet technologies enable and maintain the transmission of messages.

One of the costs of this approach is that the two case studies are generated using different methods, suggesting it might be difficult to compare the results.

Two points can be made about this. First, any case-study-based comparison of communication before and after the rise to mass use of internet technologies will face this problem because moving prior to computer-mediated communication, and the telegraph, means studying phenomena from different eras. Such a comparison is nearly certain to need different techniques to grasp the full communicative contexts because such contexts are about more than the content of messages. What needs to be studied is how the content of messages is created and moved and how receivers and senders are authorized, identified and maintained. This means paying attention to the materiality of communication in order to grasp how transmission is created and maintained, rather than focusing on the meaning of what was transmitted. This kind of study necessarily means mixing case studies researched through different methods because of a need to study phenomena from different times and spaces. Second, the comparison is of communicative practices and as long as the methods employed produce a good analysis of the nature of communication in each particular time and space, then it should be possible to compare them as such practices. That is, if the case studies produce an understanding of how communicative practices are created and maintained then it should be possible to compare these understandings, even if different methods have had to be employed to generate results.

A way forward to testing the hypothesis by retelling it in ever greater theoretical and empirical complexity is now clear. First, a more detailed theorization is required to make sense of the concepts used in the hypothesis of communicative practice. Identity, body, message, authority, legitimacy and so on all require conceptualization (see Chapter 2). This theorization will hinge on connecting the fields of communication studies and cultural studies which, while not complete strangers to each other, are also often not as closely connected as they could be. Second, such conceptualization must be done in relation to material practices, while this does not mean theory and practices must be written about simultaneously, it does mean theory cannot remain abstract while neither can material practices be understood atheoretically. Two case studies informed by the theorized hypothesis of communicative practices and engaged with the materiality of communication will take up this task. One case study focuses on pre-internet communication through a study of letters to colonial Australia in the early nineteenth century (see Chapter 3), while the second case study focuses on immersion in internet-dependent communication through an ethnography of multiplayer online games in the early twenty-first century (see Chapter 4). In both case studies, the aim is to generate as clear as possible a view of communication in their times. Finally, the results of these analyses can be reflected on to produce insights consequential to understanding communication, culture and society in the twenty-first century, particularly in the embedding of a form of communicative anxiety deeply within cultural practice (see Chapter 5).

As this study unfolds, it will become ever clearer that the state of being 'after the internet' is not being without the internet or, in a circular sense, returning to something prior to the internet, but of constantly negotiating and dealing with two broad communicative practices; one that existed prior to and that continues to exist after the rise of the internet and the other that is dependent on and could not exist without the internet. The points of intersection, whether of contradiction, unimportance or reinforcement of each other, will need to be teased out by telling the story of two communicative practices whose coexistence constitutes being after the internet.

Note

1 Hacking is a term that refers to a number of things; for this example, it is being taken as referring to cracking. For a full examination of its various meanings and their interrelations, see Jordan, 2008.

2

Communicative Practices

Introduction

The rise of internet technologies is the primary marker of new mass ways of communicating. That is, across various societies and cultures there are, inextricably intertwined with the use of internet technologies, new moments of communication enacted in the everyday that accumulate into and are enabled by wider collective practices. These new everyday moments and collective practices exist and are interrelated with pre-existing forms of communication. The key issues that need to be conceptualized then are defined by two related questions: What is communication in the everyday? And, what does it mean to conduct and repeat, that is to practice, communication?

To answer this question, it will be important to draw on both communications theory and cultural theory. To begin with the everyday-moment existing theories of communication as transmission will be touched on, and this will make it possible to contrast the everyday and transmission with ideas about communicative practices. Such a combination leaves problems, as, within communications theory, the everyday and practices have sometimes been opposed to each other as visions of communication rather than linked together to form communication. Following this, Milne's work on emails, letters and postcards introduces the idea of presence in communication to provide an understanding of what must be established for transmission to become possible. Milne begins to connect cultural studies and communication theory on the issue of presence, and this is developed subsequently through the work of Heidegger, Levinas and Derrida. Further complications emerge in the need to conceptualize both materiality, drawing on social studies of science, and performativity, relying on work by Butler and again Derrida, to fully conceptualize communication. In summary, it will be argued that communicative practices refer to the ways in which the possibility of the everyday transmission of messages is based on the creation of presence through materially enacted performatives.

Transmission and the problem of communication

Though there are a range of different understandings of communication, the term 'communicative practices' emerged within communication studies partly as a response to the conceptualization of communication as the everyday moment of passing a message from A to B (Craig, 1999, p. 7; Peters, 1999, pp. 10–29). Views such as the latter are centred on what has been called the transmission model of communication, that is often exemplified by the mathematical model of communication of Shannon and Weaver. At its simplest, their model is one in which a message is emitted from a point A, passes through the possibility of distortion or noise and is then received at point B. Transmission involves a sender, receiver, message and means of messaging, with the core problem being noise that interferes with the message. This definition of a general communication system has a notable feature.

> The fundamental problem of communication is that of reproducing at one point either exactly or appropriately a message selected at another point. Frequently the messages have *meaning*; that is they refer to or are correlated according to some system with certain physical or conceptual entities. These semantic aspects of communication are irrelevant to the engineering problem. The significant aspect is that the actual message is one *selected from a set* of possible messages. (Shannon, 1948, p. 379)

The envelope does not care what is written on the letter inside, the waves that carry radio sounds do not care if it is music or talk and, perhaps most strikingly, the internet carries digital packets which can increasingly be any type of media object as increasingly all media objects can be digitized. Dropping concern for the nature of the message in favour of the method of messaging is key to transmission models of communication in which a message A is passed from B to C. Transmission also corresponds to an intuitive or common-sense view of communication because it most closely reflects how communication is experienced in the everyday at individual instants when the event of communication happens. A letter is received, an email opened, a phone call answered, a text sent; in all such cases, the instant in which such communicative action happens usually appears to participants to be one in which a message is received having been sent (Carey, 2009, p. 33).

Yet, despite such a seemingly intuitive starting point for understanding communication, a range of objections have been made to transmission models. From within communication theory, it has been argued that transmission models are mechanical, solipsistic and ahistorical (Craig, 1999, p. 7). Outside of communication theory, the flaws of the common sense model of communication have also been pointed out. Both Luhman and Derrida, from rather different theoretical contexts, articulate the transmission model in order to critique it. Luhman pointed out three assumptions within such models: (1) that transmission assumes the message is lost to the sender and gained by the receiver, (2) that transmission implies a hierarchy in which the sender chooses what the receiver gains and (3) that transmission assumes the self-identity of the message which stays the same between sender and receiver. All three assumptions are false Luhman argues, pointing towards communication as a process in which messages, in sending, travel and reception, are negotiated within social and cultural norms (Luhman, 1995, pp. 139–41). Derrida only rarely examines the concept of communication directly, though much of his extensive work on language implies communication. Where he does directly consider communication, he takes up the transmission model, arguing that the model is inscribed within semiotics but that transmission has to presume the existence of objects and subjects and that these are affected by processes of difference, meaning they are not straightforwardly stable. If difference is necessary to these elements of a transmission model then this requires communication to have a context which clarifies and stabilizes the equivocations difference produces, and this context must both be what contains transmission and what a focus on transmission tends to conceal (Derrida, 1981, pp. 23–8; 1982a, pp. 209–10).

One of the most extensive and influential analyses of transmission as a model of communication is found in Carey's work. Carey argues that there are two models of communication that have dominated conceptualization of communication in the United States since the nineteenth century; the transmission model and the ritual model (Carey, 2009, p. 12). Carey's view of the nature transmission is the same as already discussed, effectively the passing of message A from B to C. He argues there is a complementary ritual model of communication 'directed not toward the extension of messages in space but toward the maintenance of society in time; not the act of imparting information but the representation of shared beliefs' (Carey, 2009, p. 15). Communication as ritual is here complementary to transmission, and Carey does not dismiss or refute the transmission model; rather, he argues that transmission has taken over views of communication even though it only represents a partial view. This opens the possibility that transmission should be viewed as occurring within rituals. 'A ritual view does not exclude the processes of information transmission. . . . It merely contends that one cannot understand these processes aright except insofar as they are cast within an essentially ritualistic view of communication and social order' (Carey, 2009, p. 17).

Carey argues that the reason why transmission has become the dominant and common-sense view of communication, despite being flawed, is a historical one, and here he makes a second fundamental contribution to communication studies in his argument about the significance of the telegraph. Carey argues that widespread use of the telegraph was the moment when communication was separated from transportation because the message could travel faster than the transport of physical goods. This meant that such messages could control and model the transportation of physical goods both unifying geography, for example in such things as the formation of time zones or the unification of railroad timetables, and creating virtual economies in the separation of the message from its referent, for example in the formation of futures markets (Carey, 2009, pp. 164–5 and pp. 170–1). Carey argues that during the twentieth century, because of these developments, transmission became the dominant understanding of communication, eclipsing a ritualistic view that 'conceives communication as a process through which a shared culture is created, modified and transformed' (Carey, 2009, p. 33).

Carey's views have been influential within the history of ideas about communication and have given rise to further developments in what is meant by a ritual view of communication (McCarthy et al., 2005). This development has mainly occurred in what communications theory means by the term 'communicative practices' (Craig, 1999, pp. 124–8). However, several theorists of communicative practices do not agree with Carey about transmission and instead develop critiques that question transmission's coherence paralleling Luhman and Derrida's points, for example in Deetz and Peters work (Peters, 1993; Deetz, 1994). This work provides an understanding of communicative practices developing the idea that communication is a means by which transmission is constructed in culture and society rather than seeing practice and transmission as developing together. The two innovations of separating transmission of messages from a shared culture of communication and outlining the historic effects of separating transport from communication emphasized within communication studies a sense of the importance of the creation of communication through shared social and cultural practices. Carey's work leads in this way to a shift of communication theory away from transmission and towards a broad sense of social and cultural theory.

Theories of communicative practice that follow in this tradition then, quite naturally, tend to emphasize the cultural, as the key point being made is that communication is a cultural practice and not a transmission of an object between independent subjects. However, this tends to lose the specificity of communicative practices which begin to fade into general cultural and social actions, which then underpin how in particular historical circumstances communication practices work. Communication as a specific object of analysis begins to be lost because such theories are effectively equivalent to general social and cultural theories. Peters offers an influential tracing of theories of communication that draws on Carey and generates a similar issue. Peters repeats the role of the telegraph in communication in a way that implies the loss of the material from communication: 'In principle the coefficient for signals – but not bodies – was reduced to zero, even though access and cost kept the telegraph from being the utopia of universal contact' (Peters, 1999, p. 139). This attitude is paradoxical in both Peters and

Carey because at other points they are clearly aware of the importance of technologies. Peters also considers the challenge to cultural practices of the radio.

> How to compensate for the fact that people could be in touch without appearing 'in person' was the acute question in the early history of radio and its development into a huge commercial entertainment empire. New forms of authenticity, intimacy, and touch not based on immediate physical presence had to be found. (Peters, 1999, p. 214)

But as we will see in Milne, and as subsequent chapters in this account will show, such techniques were already in existence in the ways the materialities of letters were negotiated as communicative practices. In the techniques of seals, signatures and the descriptions of bodies and selves in letters, there were many existing techniques for authenticity, touch and intimacy that were not based on physical presence. This is paradoxical as Peters is clearly aware of letters and discusses them in the context of whether a message arrives or not (Peters, 1999, p. 151), but having developed Carey's theory of the telegraphic break, that tends to emphasize the cultures of communicative practices, he fails to address the technologies of letters that made possible letter writing as communication at a distance and had done so for several hundred years prior to the telegraph. The result is then not a dismissal of technology but lessening attention paid to the everyday materiality of communicative practices, which then directs attention away from communicative technologies and towards broader cultural and social rituals.

The result is a tendency for theories of communicative practice to become more like general theories of social or cultural relations rather than being specifically theories of communication. Craig is an example of this process of shifting from transmission, to shared cultures, to communicative practices and finally to social and cultural practices.

> A stance on communication as a practice is not opposed to using communication techniques or trying to improve communication outcomes by applying scientific theory and research. Those are perfectly legitimate practices. However, it is important to see those practices within a bigger picture. Contrary to a narrow scientific-technological view, practice involves

> much more than using conscious techniques to achieve predetermined goals. The discourse about a practice is fundamentally normative, fundamentally about defining elements that constitute the practice, coordinating and regulating activities, deciding what goals are important, making evaluative judgements, and the like. (Craig, 2006, p. 41)

Communication theory in such accounts is pulled strongly towards more general social and cultural theory, leaving a gap related to how to analyse communication. Communication studies then becomes hard to distinguish from cultural and social theory. It is the refocusing away from transmission that seems to lead to a loss of focus on communication in favour of social rituals. This suggests a need to integrate transmission in some form as part of communication analysis even though this means understanding how its contradictory elements, as identified by Luhman, Derrida, Carey and others, may be maintained.

Understanding communication from the present arguments involves taking forward the insights that communicative practices are material, most obviously in their technologies, that they are in some senses the production of shared cultures and that transmission of messages involves the contradictions of transmission that need to be maintained by communicative practices. There are then at least two theoretical issues that come into focus: one need is to theorize materiality within communicative practices to ensure communication is specifically addressed, while the second need is to understand a specifically communicative sense of shared cultural practice which also shows how transmission can occur. We can begin with the latter based on a recent articulation of what communication as a practice means in the context of comparing internet- and non-internet-based forms of communication in Milne's substantial and path-breaking study comparing letters, postcards and email.

Presence one: Milne, Derrida and self-presence

Milne's study analyses forms of communication in nineteenth-century letters, late nineteenth- and early twentieth-century postcards and an early twenty-first-century email-based discussion list. Milne's study is rich in empirical detail of communication that is at a physical distance providing insights into the nature of communication across pre- and post-internet communication. Based on this empirical detail, her conclusion helps identify one component of communication consonant with arguments in the previous section.

> the fantasy of, and desire for, presence is a key element in the exchanges, communications and performances enabled by these three systems . . . the material systems of letters, postcards and email enable complex exchanges, experienced as the intimate presence of a correspondent whose corporeal body remains invisible. (Milne, 2010, p. 202)

Milne argues that presence is a common element across the different forms of communication that use the technologies of letter, postcard and email list. She draws presence from Derrida's work while giving it a particularly communicative focus: 'Presence is a term that need not refer always to material, corporeal presence. Rather, presence is an effect achieved in communication . . . when interlocutors imagine the psychological, or sometimes, physical presence of the other' (Milne, 2010, p. 2). For Milne, the overall point is that something like presence cannot be assumed but is produced or created and that across the three technologies – disparate as they are in historical specificity and everyday practices – each constructs a particular form of presence.

Peters recognizes the importance of such a view on communication, defining it historically as a school of thought associated with Heidegger which Peters argues is based on a view of 'Communication as the revelation of being to itself through language' (Peters, 1999, p. 17). Peters offers a more pragmatic and historically situated approach to communication, as he makes clear when addressing arguments of Derrida's about presence and the priority given to speech over writing:

> With his war on 'the metaphysics of presence,' Derrida is right to combat the philosophical principle that behind every word is a voice and behind every voice an intending soul that gives it meaning. But to think of the longing for the presence of other people as a kind of metaphysical mistake is nuts. (Peters, 1999, p. 270)

Whether Derrida's thought is 'nuts' is thankfully beyond this text, what is relevant in Peters is another angle on Milne's empirically based assertion of presence as the key to seeing communication in emails, postcards and letters. Peters' pragmatic view of presence confirms both its fragility and failures and its necessity. 'Because we can share our mortal time and touch only with some and not all, presence becomes the closest thing there is to a guarantee of a bridge across the chasm. In this we directly face the holiness and wretchedness of our finitude' (Peters, 1999, p. 271). Peters' view is also useful as he articulates it beyond a human-centred vision, noting that 'Communication is something we share with animals and computers, extra-terrestrials and angels' (Peters, 1999, p. 227). He offers examples in the Turing Test, Wittgenstein's lion (who if it could talk we could not understand it) and the SETI Project's search for extraterrestrial life to demonstrate how we are in communication with, in the sense of practicing a form of presence with, these non-humans, and that if we place the human at the centre of such analysis we are making a political choice (Peters, 1999, p. 230). In his care to include the non-human and the recognition of the political choice that is made if the human is implicitly or explicitly taken as the centre of analysis, and hence as the measure of value, Peters both draws on and extends other influential post-human and non-human analysts such as Haraway (Haraway, 2008, 1991).

Peters and Milne can both be understood as asserting that communication depends on the fallible creation of presence. Understanding this means exploring the nature of presence further which can be done most obviously by looking at the theorist Milne who primarily draws on Derrida and Peters who contrasts with Derrida. The key elements of Derrida's arguments in this context can be found in his analysis of the priority of speech over writing in Husserl's work, a priority Derrida considers foundational of Western metaphysics. Derrida, early on in his analysis of Husserl, introduces presence in relation to repetition. 'In order that the possibility of this repetition may be open, ideally to infinity, one ideal form must assure this unity of the indefinitely and the idea: this is the present, or rather the presence of the living present' (Derrida, 1973 p. 6). And one type of living present forms presence, that is forms the means by which the stability of meaning is known, and this is speech. The quality of speech that allows it to play this role is that the meaning uttered is present to the utterer in the utterance.

> The 'apparent transcendence' of the voice thus results from the fact that the signified, which is always ideal by essence, the 'expressed' *Bedeutung*, is immediately present in the act of expression. This immediate presence results from the fact that the phenomenological 'body' of the signifier seems to fade away at the very moment it is produced; it seems already to belong to the element of ideality. . . . This effacement of the sensible body and its exteriority is *for consciousness* the very form of the immediate presence of the signified. (Derrida, 1973, p. 77)

Derrida locates Husserl's prioritization of speech in this absolute sense of self-presence because 'the speaking subject hears himself *[sic]* in the present' (Derrida, 1973, p. 78; 1976, pp. 7–8). Derrida then points out that even here difference and repetition are at play, being constituent to the very division that separates the self-presence of speech from the non-self-presence of writing.

> Even while repressing difference by assigning it to the exteriority of the signifiers, Husserl could not fail to recognize its work at the origin of sense and presence. Taking auto-affection as the exercise of the voice, auto-affection supposed that a pure difference comes to divide self-presence. In this pure difference is rooted the possibility of everything we think we can exclude from auto-affection: space, the outside, the world, the body, etc. As soon as it is admitted that auto-affection is the condition for self-presence, no pure transcendental reduction is possible. But it was necessary to pass through the transcendental reduction in order to grasp this difference in what is closest to it – which cannot mean grasping it in its identity, its purity, or its origin, for it has none. We come closest to it in the movement of differance. (Derrida, 1973, p. 82)

Differ*a*nce is Derrida's combination of 'differing' and 'deferring' that turns out to be the foundational moment underpinning Husserl's logic. While Husserl, Derrida argues, felt he had reached absolute self-presence in speech, this self-presence is itself produced by differing and deferring that cannot be in the auto-affection of speech and so is outside presence understood as speech. Accordingly, what alone is inevitable is differ*a*nce.

Derrida extends and complicates this argument through a range of interventions. His detailed essay on differ*a*nce notes that the turn he identifies is found in the work of Heidegger, Nietzsche and Freud in which presence turns from being a power to synthesize and instantly reassemble meaning into an effect of a system that is itself based on differance (Derrida, 1982b, pp. 16–17) and states in related work: 'The system of "hearing (understanding)-oneself-speak" through the phonic substance – which *presents itself* as the nonexterior, nonmundane, therefore nonempirical or noncontingent signifier – has necessarily dominated the history of the world during an entire epoch' (Derrida, 1976, pp. 7–8). Such a framing of this period in Derrida's work gives it much of its millennial or revolutionary feel with its seeming war on the domination of writing by speech and the implied high stakes that involve revolutionizing Western metaphysics.

Derrida's analysis defines the notion of presence within the debate about whether speech produces a privileged type of presence – in which the self is transparently present to the self – and in exposing how speech as presence is itself produced on the basis of other factors, he also locates presence as occurring within certain contexts, both historical and philosophical. This is the key lesson Milne draws and which she argues her evidence confirms, 'As Derrida reminds us, "presence" is not an empirical neutral fact. Rather, presence is an effect generated through various representational systems by the desires shaped by specific cultural and philosophical ideologies' (Milne, 2010, p. 72). For example, Milne examines the way letter writers represent their bodies in the text of their letters, arguing that a key element of epistolary practices in the nineteenth century was the written production of a body in letters which helped construct presence between the communicants. Milne argues this representation of the writer's body, which remained materially distant, was important enough that actually meeting and so confronting the imagined body with the physical body might destroy the presence that had been developed in letters.

> The sense of presence they experience in letters may be destroyed by a physical meeting . . . when the materiality of the epistolary system is used as a metonym for the corporeal body . . . references to the corporeal body often create a sense of 'presence,' in the Derridean understanding of this concept.

> The body is used to signify a presence that exists outside of, or is in excess of, the realm of the physical encounter. Arguably, this is a defining feature of the familiar letter: its ability to convey virtual presence by using signs of the material and corporeal. (Milne, 2010, p. 90)

Presence in the context of communication is then a product of material and cultural practices. From these arguments drawn from Derrida and Milne, I can begin to define communicative practices as the ways and means of creating presence outside of the presumption that senders and receivers of messages are automatically present in the message's utterance. The core elements of Derrida that Milne draws on are in a sense a negative argument by which any sense of absolute self-presence is disturbed and revealed as, for Derrida, dependent on differ*a*nce, and, for Milne, dependent on differ*a*nce as engaged by various ideologies and technologies. Accepting this, we still need to ask if presence is created or constructed, as Milne shows so well in her analyses of letters, postcards and emails, and if Derrida makes it clear there is no obvious 'self-presence,' then what is it that is being created? If presence is not a self-evident and transparent presence of the self to the self then what is a created or constructed communicative presence?

Presence two: Heidegger, Levinas and face-to-face

If presence is not transparent and absolute self-presence that founds an autonomous and self-consistent subject, then Milne's suggestion of a communicative presence needs to be expanded and understood. This raises a question of being and Being that is within presence and communication because this questions the nature of the self. Derrida argued that to follow the critique of presence, which Milne draws on, the path is through Heideggerian philosophy (Derrida, 1976, p. 23).[1] The point is not that Heidegger's thought is the end of the theory of Being, but rather that it is essential to reach an understanding of presence 'by way of' Heidegger's thought. Having picked up the trail based on Milne's work, leading to Derrida we can now continue down this path, taking up Derrida's pointer that Heidegger is a necessary interlocutor.

Peters also points towards Heidegger who, he argues, creates a sense of communication which does not see communication as the openness of one mind to another, but as part of our fundamental 'being-with' others. Peters says of Heidegger:

> To be human is to be linguistic and social. Speech can make our relations explicit, but there is no question for Heidegger of communication's failing between people any more than there is of people's ceasing to dwell in societies and in language. We are bound together in existential and lived ways before we even open our mouths to speak. Communication here does not involve transmitting information about one's intentionality; rather, it entails bearing oneself in such a way that one is open to hearing the other's otherness. (Peters, 1999, p. 16)

Peters here invokes the idea of 'being-with' that is always already there. This also invokes the core Heideggerian concept of Dasein or Being. It should be noted that the work of Heidegger and work on Heidegger are vast, complex and at times troubling but, for the purposes of this account, it is important to follow the thread that is relevant to communication and this has led to being-with and Dasein, which will here be drawn from their formulation in *Being and Time.*[2]

Heidegger's notion of Being is, from the start, Being in the everyday and he argues Being as experienced by beings has been passed over often precisely because it has this 'average everydayness' (Heidegger, 1980, p. 69). Accordingly as he begins to uncover the form of Dasein, Heidegger begins with beings in ordinary interactions; for example in buying a book, in walking around a field, looking at someone else's boat or discussing work (Heidegger, 1980, pp. 153–4). A number of things are important about this, not only does Heidegger start with the way Being works in everyday interactions of beings, he also shows this means that there is no outside to beings.

> In clarifying Being-in-the-world we have shown that a bare subject without a world never 'is' proximally, nor is it ever given. And so in the end an isolated 'I' without Others is just as far from being proximally given. If, however, 'the Others' already *are there with us* in Being-in-the-world . . . even this should not mislead us into supposing that the *ontological* structure of what is thus 'given' is obvious, requiring no investigation. (Heidegger, 1980, p. 152)

Our being-there in the world is always as part of interactions with others, often simple and habitual, but this is only the beginning of necessary analysis, Heidegger argues, and does not reveal the nature of Being in being-in-the-world or in being-there. For this Heidegger begins to explore the nature of Dasein.

The result, for Heidegger, of recognizing that beings are always in the world related to others is that for each of us our Being is always 'one's own,' or 'has mineness'; Dasein is personal, while at the same time we always encounter others engaged in their kind of 'being-in-the-world'. Being-with is thus integral, it is an 'existential characteristic of Dasein' (Heidegger, 1980, pp. 156–7). 'Not only is Being towards Others an autonomous irreducible relationship of Being: this relationship, as Being-with, is one which, with Dasein's Being, already is' (Heidegger, 1980, p. 162). The Being of beings is a relationship that always already exists between beings and their others, expressed generally as between Self and Other. Dreyfus' account of Being-with identifies this point in Heidegger.

> Heidegger's basic point is that the background familiarity that underlies all coping and all intentional states is not a plurality of subjective belief systems including mutual beliefs about each others' beliefs, but rather an agreement in ways of acting and judging into which human beings, by the time they have Dasein in them, are 'always already' socialized. Such agreement is not conscious thematic agreement but is prior to and presupposed by the intentionalistic sort of agreement arrived at between subjects. (Dreyfus, 1995, p. 144)

Being is accordingly about various ways of being-with others. However, Heidegger points out that empathy with others is not a 'primordial existential phenomenon, any more than knowing in general' (Heidegger, 1980, p. 163) because empathy only arises consequent on one's own Dasein already being-with others. The gap opened up here is that certain primordial existential phenomena, so understood because they are part of Being-with-one-another and Dasein, mislead from empathy.

> Our analysis has shown that Being-with is an existential constituent of Being-in-the-world. Dasein-with has proved to be a kind of Being which entities encountered within-the-world have as their own. So far as Dasein *is* at all,

> it has Being-with-another as its kind of Being. This cannot be conceived as a summative result of the occurrence of several 'subjects'. Even to come across a number of 'subjects' becomes possible only if the Others who are concerned proximally in their Dasein-with are treated merely as 'numerals'. (Heidegger, 1980, p. 163)

The others are encountered not as a full beings-with-and-towards-one-another because Heidegger understands knowing and empathy as appearing after the encounter with the Other. Prior to knowing and empathy, the Other (others) are encountered numerically, as a series of others with which our or mine Dasein cannot encounter as full being-with. Authentic Dasein is here submerged or absorbed into the world of its concerns, its everyday concerns.[3]

This diversion from authentic Dasein introduces both the quest to recover Dasein and the main barrier to such a recovery in Heidegger's account of the 'they' or *das Man*. By this Heidegger refers to the Other who is not a definite other but others as conglomerated and represented within the generality of Other. Each specific other, who could or should give rise to being-with, becomes a generalized and equalized other, such that one's Dasein, one's mineness, becomes part of the generalized 'they' that distances ourselves from being-with (Heidegger, 1980, pp. 163–5). The 'they' of this conglomerated Other is a dictatorship of the average and a drive to reduce all to what is publicly part of each other; for example, we dress either with or against a fashion which is driven by 'they' but in either case we are defined in fashion terms by this 'they'.

> Overnight, everything that is primordial gets glossed over as something that has long been well known. Everything gained by a struggle becomes just something to be manipulated. Every secret loses its force. This care of averageness reveals in turn an essential tendency of Dasein which we call the 'levelling down' of all possibilities of Being. (Heidegger, 1980, p. 165)

Heidegger has now defined the quest to reveal Being or to recover the revelation of Being in authentic Dasein that he has already realized but which the 'ordinary' folk, the dictatorship of the average, prevent realization of in the everyday. Having found the fundamental nature of existence to be 'being-with,' Heidegger's argument for the concealment of this turns on his rejection of others as inauthentic and mired in a reduction of all to the average. From finding the ordinary and mundane everyday at the core of Being as multiple, differential relations of being-with, Heidegger argues this seemingly similar mundane conceals the revelation of Being's meaning.

Yet this seems odd and I can ask, in relation to communication and presence, why is the everyday both the meaning of Being and the barrier to realization of this meaning? It does not seem immediately clear that ordinary activities such as buying a book or sitting at a train station are necessarily routines that demand we subject ourselves to the average in each other. Peters and Milne have already argued that in regard, at least to communication, the everyday may be fragile but it is certainly capable of sustaining communication and that this itself points to the existence of presence and being-with. The point they make is not of a fundamental failure of the possibility of selves and others being-with in communication, but their work demonstrates such communication occurs and infer from this that there must be presence and being-with, even if fallible, leaving the question of how such presence is theorized. Such accounts do not seem to square easily with Heidegger's in which the everyday is the arena of failed being-with but rather accord more closely with an account like Haraway's focus on the entanglements that make up each being-with. Here Haraway considers Heidegger's means of freeing Dasein to unconceal the truth.

> To achieve this great voiding of illusion, to grasp 'negativity', to be free, to understand one's captivity rather than merely live it as an animal, . . . a man in Heidegger's story allows the terrible experience of profound boredom to drench his whole self. . . . Only from this great destroying and liberating antiteleological negativity, this perfect indifference, can Dasein . . ., true human being, emerge. Only from this 'open' can man grasp the world with passion, not as stock and resource, but in unconcealment and disclosure freed from technique and function. My 'open' is quite other, if similarly lustful for nonteleological understanding. It emerges from the shock of 'getting it': *This* and *here* are who and where we are? What is to be done? How can respect and response flourish in *this* here and *this* we, even as this *we* is the fruit of entanglement? . . . Never certain, never guaranteed, the 'open' for companion species becomes possible in the contact zones and unruly edges. (Haraway, 2008, p. 367–8)

If Heidegger's mistrust of the masses is questioned then it is possible to refuse to see Being as concealed in the way Heidegger argues it must be. As Haraway suggests, we can exist within the unruly and not see these as necessarily concealing the truth of existence. This means that while Heidegger establishes presence as forms of being-with, to provide a specifically communicative sense of this presence, it is worth examining a further theoretical resource that is consonant with the idea of being-with but, following Haraway, is dissonant to Heidegger on the question of betrayal through the average everydayness of the Other. This can be found in Levinas' work which also revolves around a key concept for internet and non-internet communication in what it means to be 'face-to-face'.

What do we mean when we say we are face-to-face? The question of co-presence may be contained in that question. For Levinas, the face we see is the face of someone else, someone other than us, an other, an other's face. The general conditions for understanding what significance a face has will involve understanding the Other; that which convinces us we are not the sole entity and centre of all worlds, that we are not unproblematically the same. Such an encounter is the inescapable assumption we must make in order to reach an understanding of the face. If we argue there are no others, then the selves can see no face and there can be no face-to-face. This accords with the idea of being-with in Being. Thus, as we refuse to presume that we know anything except that there is a face-to-face we are led to consider the general terms of Self and Other, with no presumptions concerning distance, the physical or bodies. We can immediately see how this is similar to Heidegger's being-with but also different in its assumption of what happens in the encounters of selves and others. Levinas offers an analysis of the face, that both draws on the Heidegger while also specifying presence differently.

> The way in which the other presents himself, exceeding *the idea of the other in me*, we here name face. . . . The face of the Other at each moment destroys and overflows the plastic image it leaves me, the idea existing to my own measure and to the measure of its *ideatum* – the adequate idea. It does not manifest itself by these qualities. . . . It *expresses itself*. The face brings a notion of truth which . . . is not the disclosure of an impersonal Neuter, but expression: the existent breaks through all the envelopings and generalities

> of Being to spread out its 'form' the totality of its 'content', finally abolishing the distinction between form and content. (Levinas, 1969, pp. 50–1)

What does Levinas see in a face? He sees the Other, the fact that the Self – myself – is not the only existent, not the only existing thing meaning the world, my world, cannot revolve solely around my desires, conceptions and understanding; 'the *phenomenon* that is the apparition of the Other is also *face*' (Levinas, 2006, p. 31). This breaking up, this moment in which any self realizes it is caught up in the Self and the Other is foundational.

> What is exceptional . . . is that I am ordered toward the face of the other. . . . All the negative attributes which state what is beyond the essence become positive in responsibility, a response answering to a non-thematizable provocation thus a non-vocation, a trauma. This response answers, before any understanding, for a debt contracted before any freedom and before any consciousness and any present, but it does answer, as though the invisible that bypasses the present left a trace by the very fact of bypassing the present. That trace lights up the face of a neighbour, ambiguously him *before whom* (or *to whom*, without any paternalism) and him *for whom* I answer. (Levinas, 1981 pp. 11–12)

Face is the presence of the Other to the Self and of founding Being in the inescapable recognition of other beings. This is not far distant from Heideggerian being-with but construes the Self and Other encounter differently, seeing the multiple relations of selves and others that make up Self and Other but not mistrusting all the others. It is in this way a rejection of Heideggerian mistrust and an opening to Harawayian joy, entanglement and encounter. But what is the relationship that each being experiences in feeling or finding others and which is constitutive of Being in the general relationship of the Self to the Other? When I say 'experience,' it is not an experience in any everyday sense of the word, it is the experience on which everydays may be founded. Levinas argues it cannot be anything but a primary and inescapable responsibility, such that he asks, 'But isn't this servitude? *Not being able* to get out of responsibility?' (Levinas, 2006, p. 52), and he considers the hostage to be a profound understanding of this relationship in which responsibility for others and to others is always already there.

The hostage remains an important implication of Being in which the Self is understood as primordially responsible to the Other, but it is only one model. Levinas also terms this relationship a conversation. The conversation does not contradict the hostage as an expression of Being but instead interprets it differently, away somewhat from responsibility, the implication of the interaction of selves and others, and towards the interaction itself.

> Conversation, from the very fact that it maintains the distance between me and the Other, the radical separation asserted in transcendence which prevents the constitution of totality, cannot renounce the egoism of its existence; but the very fact of being in a conversation consists in recognizing in the Other a *right* over this egoism, and hence in justifying oneself. (Levinas, 1969, p. 40)

Being face-to-face can thus be understood as a fundamental expression of existence, of what is inescapable about existing. This is the relationships of selves to others which cannot be escaped, avoided or denied, whether expressed in the intense intimacy of mother-child or the multifarious encounters of the everyday. Returning again to Haraway, instead of boredom disclosing authenticity we find play in these intermingled being-withs. Play which 'also requires something . . . namely, *joy* in the sheer doing. . . . Play makes an opening. Play proposes. . . . Play is not making a living; it discloses living' (Haraway, 2008, p. 240). This is not play as just something trivial but play as also revealing what it is to live. Being is here constituted in the serious play of relations. Being is both the conversation and a relationship of hostage between Self and Other and both express the deepest and inescapable responsibility of beings to each other. It is a change of being-with to becoming-with that, I would argue, is substantially similar to Levinas' name for this relationship as the Face and to be face-to-face is to encounter the Other.

> A relation whose terms do not form a totality can hence be produced within the general economy of being only as proceeding from the I to the others, as a *face to face*, as delineating a distance in depth – that of the conversation, of goodness, of Desire – irreducible to the distance the synthetic activity of the understanding establishes between the diverse terms, other with respect to one another, that lend themselves to its synoptic operation. (Levinas, 1969, p. 39)

Being face-to-face in Levinas' thinking is a fundamental moment in which the self is forced to extend beyond itself and, in that extension, to realize the existence of many others leading to the realization of the Other. Levinas argues that this realization of the necessity of the Other's face, of the reality of living beyond the Self, because it extends a being beyond itself, takes the form of a conversation and a hostage taking, and that this means it is a relationship of responsibility.

While Levinas' concern is with fundamental ontological questions, my arguments are looking for something a little less fundamental in an understanding of communicative presence, which can now be summarized drawing on the preceding arguments. Both Levinas and Heidegger take up a sense of being-with. However, where Heidegger argues there is an immediate loss of presence in the midst of the average masses, Levinas sees responsibility and this is given a communicative direction by being interpreted as both conversation and hostage. It is in this sense that it becomes closer to Haraway's idea of becoming-with. It is this sense of being-with/becoming-with as conversation that can be taken forward to understand presence in communication, based on its two-edged sense of both care and capture. This does not go as far as Levinas does in claiming a fundamental moment of Being, but it asserts the idea that presence in communication is the unstable, difference-dependent creation of being-with/becoming-with or the face-to-face, whose establishment in specific spaces and times makes transmission of messages possible. Building on Levinas and Haraway helps to define presence in communication means as the construction or creation or recognition of a face-to-face in which selves and others are, hostage-like, related in a conversation and play. Building on Heidegger, this conversation is made in the average everydayness of social life. Building on Derrida, this conversation engages differ*a*nce and the necessary instability and performative nature of relations that differ*a*nce entails. Finally, building on Milne and the prior work of communication theorists like Carey and Peters, communication exists when there is presence which enables transmission.

This means that for communication to happen there must be presence. Further, this means that though there will be a form of presence within conditions of differ*a*nce and everydayness in any form of communication, the nature of this presence and its conditions cannot be known outside of specific and particular communicative forms. To understand a spatially and historically specific form of communication, the way a particular face-to-face as presence is created and maintained within conditions of differ*a*nce and everydayness will need to be specified. This presence then opens the possibility of transmission. Each such understanding of a particular form of communicative presence and the transmission it enables is what is meant in the following by a communicative practice. Put another way, a communicative practice is a historically and spatially specific way of constructing presence and enabling transmission. Knowing this however still leaves two concepts to be explicated for the theory of communication presented here to be clear; performativity or practice and materiality. Though the term 'practice' and 'performative' have been used, these need to be explored to understand what communicative actions in the everyday mean, and because presence is specific in particular times and spaces, communication is something material which also needs conceptualization. Performativity and practice will be examined next, after which suggesting a concept of matter will complete a full conceptualization of communicative practices.

Performativity

Derrida's (contested) use of Austin's concept of the speech-act and Butler's subsequent extension and criticism of him form a way of understanding performativity and its relation to practice; this extends and connects to the points opened up by Milne's use of Derrida on presence while also adapting influential work in cultural theory on performativity. The argument reconnects to the previous discussion when Derrida analysed Husserl to claim that differ*a*nce is inescapable and opens up the instability of presence. What was outlined but not discussed in relation to differ*a*nce is the meaning of iterability within differ*a*nce. It is here that Derrida argues that repetition or iterability is a problem because repetition seeks to return the same but each return is necessarily of something different. Put another way, if it were not different in some way, then it could not be a repetition but would in fact be the same thing, but if it is different, then it cannot be a repetition. Further, in communicative practice, as so far defined, presence creates the possibility of transmission which requires repetition to be an ongoing form of communication. Communication requires iterability in which the same is repeated but only on condition of a necessary difference (Derrida, 1988; Loxley, 2007, pp. 62–87).

Butler picks up this point and notes what appears to be an opposition or contradiction in Derrida's work on differance and iterability between that which exists across different significations and the particular signification itself. She asks: 'What guarantees the permanence of this crossed and vexed relation in which the structural exceeds and opposes the semantic, and the semantic is always crossed and defeated by the structural?' (Butler, 1997, p. 150). Immediately after posing this question Butler argues that it is a question to take seriously if one wants to think through 'the logic of iterability as a social logic' (Butler, 1997, p. 150).

To see the importance of this social logic and its role in repetition and iterability, it is important to open up performativity as a particular kind of relation. Performativity is the relation in which the performance produces both the performed and the performer. Butler suggests that performativity can be understood through two closely linked sets of concepts: repetition, iteration and citation, on the one hand, and practices, on the other hand.

> If a performative provisionally succeeds . . . then it is not because an intention successfully governs the action of speech, but only because that action echoes prior actions, and *accumulates the force of authority through the repetition or citation of a prior and authoritative set of practices*. It is not simply that the speech act takes place *within* a practice, but that the act is itself a ritualized practice. (Butler, 1997, p. 51)

And the repetition and citation performs the particularly singular function of performativity: 'Indeed, is iterability or citationality not precisely this: *the operation of that metalepsis by which the subject who "cites" the performative is temporarily produced as the belated and fictive origin of the performance itself?*' (Butler, 1997, p. 49). Practices are referred by Butler to two key concepts in Austin's of the speech-act and Althusser's of interpellation. Both these conceptualize the ways in which a performative produces both the subject and object of itself, the way that a performative produces the 'effects that it names' (Butler, 1993, p. 1).

But this performativity still relies on and is troubled by the difficulty Derrida argued attends to repetition and iteration because of the impossibility of repetition. Butler builds on performativity to argue that Derrida has, in a sense, conceptualized the problem of iterability as being universal. 'Derrida appears to install the break as a structurally necessary feature of every utterance and every codifiable written mark, thus paralyzing the social analysis of forceful utterance' (Butler, 1997, p. 150). Derrida, in Butler's argument, makes it difficult to analyse the politics of communication or utterance because utterance is subject to the universal problem of iteration and repetition. For this reason, Butler argues she needs an account of the 'social iterability of the utterance' (Butler, 1997, p. 150). To understand this social iterability Butler needs to break out of a structural dominance in which every utterance is iterated and hence is suspect for philosophical reasons. She takes up ideas of social sedimentation in which repetition builds on previous repetitions that are maintained socially through performativity's ability to create performer and performed in performance.

Butler argues that to overcome the abstract problem of difference, there must be a force that is both integrated into performatives, in fact is essential to them, but at the same time is beyond their intelligibility. This she argues is the irreducibility of the speaking body; 'the abiding incongruity of the speaking body, the way in which it exceeds its interpellation, and remains uncontained by any of its acts of speech' (Butler, 1997, p. 151). It is the speaking body that causes a 'scandal' in the processes of iterability as social sedimentation and so opens up the possibility that something like the same can traverse the necessary difference.

> I would agree with Bourdieu's critique of some deconstructive positions that argue that the speech act, by virtue of its internal powers, breaks with every context from which it emerges. That is simply not the case, and it is clear to me, especially in the example of hate speech, that contexts inhere in certain speech acts in ways that are very difficult to shake. On the other hand, I would insist that the speech act, as a rite of institution, is one whose contexts are never fully determined in advance, and that the possibility for the speech

> act to take on a non-ordinary meaning, is one whose contexts are never fully determined in advance, and that the possibility for the speech act to take on a non-ordinary meaning, to function in contexts where it has not belonged, is precisely the political promise of the performative. (Butler, 1997, p. 161)

The ability to break with the sedimentation of iterations is a political promise that, for Butler, comes from the body. It is this body that Butler posits as the means by which a break can be made, against Bourdieu's dismissal of performativity.

But what is a body and a speaking body? These seem to be to Butler a way of constructing social iterability which allows performativity, but this turns on the idea of a speaking body whose nature is, in turn, unclear. Some, particularly Barad, argue that within Butler's invocation of the body there is a sense of materiality, of the body as a form of matter, and that Butler is looking to bring matter and discourse into closer proximity, but in fact does not do so. 'Questions about the material nature of discursive practices seem to hang in the air like the persistent smile of the Cheshire cat' (Barad, 2007, p. 64). Barad argues that the issue is not just that of bodies but behind bodies of the nature of matter. While Barad adopts a similar approach to performativity as Butler, she does so to open up a way beyond divisions which leave matter out of analysis. 'A performative understanding of scientific practices, for example, takes account of the fact that knowing does not come from standing at a distance and representing but rather from *a direct material engagement with the world*' (Barad, 2007, p. 49). In the present context, if communication is now understood as the creation of presence through certain performatives, to fully understand such performatives and how they overcome the problem of repetition, we need to insert matter into this and ask, what does materiality mean in such a form of communication? Moreover, Butler's assertion of the body and Barad's analysis of this assertion suggests that performativity cannot be understood fully without understanding matter and its relations to the social logics of performativity.

While Derrida and Butler allow us to understand performatives as particular ways of asserting the performer, performed and performance in the one action, they still struggle to establish the sociality or social logic of performativity as a way of limiting or regulating the effects of the necessary failure of repetition. Butler looks to help from a third concept that has also already emerged as being important to theorizing communication in matter and the body. Examining matter will allow presence and performativity to be brought together as a definition of communicative practices.

Matter and communication

As it is Barad I have drawn on to make the connection between performativity and materiality, it is worth taking up her account to help articulate a compatible understanding of matter. Barad's work explicitly attempts to put matter back into play in social and cultural theory; she is important for having enforced a head-on collision between 30 years of post-structural, constructivist science studies, performativity theory and the idea of matter. She also concisely puts the case for what is at stake in the idea of matter and why it needs to be analysed.

> It is difficult to imagine how psychic and sociohistorical forces alone would account for the production of matter. Surely it is the case – even when the focus is restricted to the materiality of 'human' bodies (and how can we stop there?) – that there are 'natural', not merely 'social', forces that matter. . . . What is needed is a robust account of the materialization of *all* bodies – 'human' and 'nonhuman' – including the agential contributions of all material forces (both 'social' and 'natural'). This will require an understanding of the relationship between discursive practices and material phenomena. (Barad, 2007, p. 66)

Barad takes forward the notion of the performative, which argues that we may make our being through practices in which produced and performer are created in the performance, by arguing that understanding of the performative has been flawed by its failure to account for the material. I have noted the importance of the body for Butler's account of the social logic of performativity and the importance of an idea of materiality within the idea of the body. Barad builds up her conceptualization of matter through an encounter with classical quantum mechanics in Niels Bohr's work. Barad interprets Bohr's work on complementarity as, in part, resulting in the conclusion that the nature of an object depends on the way it is measured, allowing some objects to have apparently contradictory properties (Barad, 2007, p. 107). Such a view is usually closely connected to Heisenberg's uncertainty principle but Barad focuses on a specific reading of Bohr's concerning complementarity.

> Bohr's argument for the indeterminable nature of measurement interactions is based on his insight that concepts *are defined by the circumstances required for their measurement*. . . . Bohr's indeterminacy principle can be stated as follows: *the values of complementary variables (such as position and momentum) are not simultaneously determinate*. The issue is not one of unknowability per se; rather, it is a question of what can be said to simultaneously exist. (Barad, 2007, pp. 108 and 118)

Barad argues that for Bohr objectivity is fixed at the point at which the measuring apparatus is fixed. One of Bohr's famous thought-experiments demonstrated that light could have either a wave or particle form depending on how it is measured, thus giving light contradictory qualities. Barad's reading of Bohr is that this contradiction never comes to exist because light is resolved as being either a wave or a particle when a particular apparatus is set up that measures a particular instance of light and in doing so objectifies it. There are, of course, considerable complexities and arguments behind this interpretation of Bohr, but for present purposes the important point is not the historical accuracy of Barad's account of Bohr but that Barad then generalizes her reading as a discursive *and* materialist theory, refusing to prioritize one over the other while asserting the necessity of both.

For Barad, the notion of the apparatus can be generalized so that it becomes the condition of any phenomena existing and not just laboratory-based experimental phenomena. With the refusal of primacy to either the discursive or material and a commitment to understanding phenomena as performed and not simply given, Barad adds the notion of the apparatus as the moment when possibilities are 'cut' or objectified.

> In my agential realist elaboration of Bohr's account, *apparatuses are the material conditions of possibility and impossibility of mattering*; they enact what matters and what is excluded from mattering. . . . the apparatus specifies an agential cut that enacts a resolution (within the phenomenon) of the semantic, as well as ontic, indeterminacy. *Hence apparatuses are boundary-making practices*. (Barad, 2007, p. 148)

Apparatuses can make cuts, on an analogy to measurements, that stabilize a particular state to which material and discursive contributions are made and remade. Matter and discourse each refer to ways of making contributions, of intra-acting, to entanglements of bodies, signs, matters, humans, animals, post-humans, all of which become a particular phenomenon out of their multifarious possibilities when the cut is made. Phenomena are then part of sedimented worlds in which particular ways of being and doing become 'normalized' by their cuts being repeated. Thus despite the seeming instability and multiplicity of Barad's account, she argues for the stability of the world we live in through the iterations of performatives that have the characteristics of apparatuses.

> Since different agential cuts materialize different phenomena . . . our intra-actions do not merely effect what we know and therefore demand an ethics of knowing; rather our intra-actions contribute to the differential mattering of the world. *Objectivity means being accountable for marks on bodies, that is, specific materializations in their differential mattering.* We are responsible for the cuts that we help enact not because we do the choosing (neither do we escape responsibility because 'we' are 'chosen' by them), but because we are an agential part of the material becoming of the universe. Cuts are agentially enacted not by willful individuals but by the larger material arrangements of which 'we' are a 'part'. (Barad, 2007, p. 178)

Barad's work pushes the material back into an accounting for bodies and the world, while at the same time retaining the power of the discursive. Her account goes further in understanding the nature of matter in the attempt to understand the body. Yet it is important to consider some difficulties that this last accounting for responsibility brings to the fore.

Despite Barad's complex account, it seems that responsibility remains caught in a duality of something like a structure (even when structure is understood as a sedimented series of cuts that stabilize multiple and unstable entanglements of matter and discourse) and something like an agent (the role we all have in a cut being made). It is striking in the quotation just given: how cuts we do not choose to make still cut and we are chosen by cuts but we are also not free from the choices we make by virtue of being entangled within a cut which chooses us. It begins to feel like the dualities that Barad has found Bohr so helpful in overcoming are here coming apart and reasserting themselves. This gives Barad's account of responsibility and agency a feeling of having contradictory views sewn together and that this sewing is itself a cut, in Barad's technical sense, and so forms the phenomena of agency and responsibility as simultaneously and always both the responsibility to make choices in 'cutting' and having been chosen by a cut.

This strain in overcoming the dualisms of so many fundamental categories of thought and practice – matter and language, ontology and epistemology, human and post-human, structure and agency – sometimes suggests that Barad's own language reintroduces these divisions. 'Intra-actions are nonarbitrary, nondeterministic, causal enactments' (Barad, 2007, p. 179). Such statements begin to appear problematic because instead of the complex but determining moment in which a cut stabilizes objects, Barad now interprets this as causes that do not determine. Similarly,

> *matter is a dynamic intra-active becoming that is implicated and enfolded in its iterative becoming. Matter(ing) is a dynamic articulation/configuration of the world.* In other words, materiality is discursive (i.e., material phenomena are inseparable from the apparatuses of bodily production; matter emerges out of, and includes as part of its being, the ongoing reconfiguring of boundaries), just as discursive practices are always already material (i.e., they are ongoing material [re]configurings of the world). (Barad, 2007, pp. 151–2)

Rather than overcoming the divisions and boundaries of a matter and discourse duality, Barad has engaged them in a process in which they are constantly entangled in multiple ways with multifarious forms of contribution.

This driving, complex attempt to push theory of matter forward and the use of an understanding of science make Barad's work both fascinating and important, yet Barad also finds the categories she seeks to overcome both must be asserted at the beginning of her work and, I argue, they resurrect themselves at the conclusion. The cost of this emerges in a prioritization of processes over categories. Everything for Barad begins as becoming and ends as becoming and it is only in such processes that all entanglements are made. This ultimately falls short for processes cannot exist or dominate the categories they require to make a process happen, this is not to suggest that categories should be dominant but to point out that Barad has strongly prioritized the means by which phenomena are produced such that what seems to matter is not what a particular phenomena 'is' but that it 'moves'. This undermines her identification of the moment when a thing, a category, is made in the cut because the cut is undermined as a moment of categorization or of stasis by the becomings that make things move. Can a cut ever be made within such ongoing processes of becoming?

Barad sets up an opposition or duality between the discursive and the material that she then seeks to reconcile, and this is the key duality that seems to me to recur within her work rather than being resolved. Perhaps this means my points are less a criticism of her logic than a disagreement with her that there is such a duality that needs resolving. The idea of the cut emerges as a powerful suggestion because it conceptualizes how even within discourses a moment may occur that objectifies. Barad tangles with contemporary accounts of matter by asserting the entanglement of the given and the word. 'Matter and meaning are not separate elements. They are inextricably fused together, and no event, no matter how energetic, can tear them asunder' (Barad, 2007, p. 3). Barad pursues matter as itself being active and how this activity has been missed or mislaid within discursively dominated accounts. However, Barad's reaching for matter seems to give way to accounts that drift back into the very categories of discourse and matter that her account of performativity was intended to move beyond.

I feel that despite Barad's account as soon as she asserts, 'It is difficult to imagine how psychic and sociohistorical forces alone would account for the production of matter' (Barad, 2007, p. 66) then there is an irreconcilable duality of matter and discourse which returns within her work and is a perennial difficulty many feel for discourse theory in the face of matter. This difficulty manifests itself in a desire for matter or a given, a desire which can only be articulated through discursive practices that themselves necessarily cannot reach matter because they are discursive. Serres registers a similar desire when pursuing something that is beyond language while recognizing the irony that language always reclaims whatever seems beyond language: 'The given I have called hard is sometimes, but not always, located on the entropic scale: it pulls your muscles, tears your skin, stings your eyes, bursts your eardrums, burns your mouth' (Serres, 2008, p. 113). Yet as soon as this desire for the given is enunciated, Serres recognizes the infuriating and inevitable failure to fulfil that desire: 'Is there a single given independent of language? If so, how do we apprehend it? The discussion is over as soon as it begins: no-one knows how to say what is given, independently of language. Any description of the aforementioned things is merely presenting data in relation to the language being employed. The thing itself flees along the infinite asymptote of utterance' (Serres, 2008, p. 112). Butler registered a related difficulty when reflecting on her analysis of the body.

> I began writing this book by trying to consider the materiality of the body only to find that the thought of materiality invariably moved me into other domains. I tried to discipline myself to stay on the subject but found that I could not fix bodies as simple objects of thought. Not only did bodies tend to indicate a world beyond themselves, but this movement beyond their own boundaries, a movement of boundary itself, appeared to be quite central to what bodies 'are'. I kept losing track of the subject. I proved resistant to discipline, Invariably, I began to consider that perhaps this resistance to fixing the subject was essential to the matter at hand. (Butler, 1993, p. ix)

Butler here opens up a way of using the performative approach to bodies and matter to move past the difficulties so far seen in understanding matter beyond discourse. Perhaps the inability to focus on matter is part of its nature. Perhaps matter is that which goes beyond our discursive abilities and so becomes unspeakable; Serres calls on stings and tears and the irony that these must be called in language, Butler finds she is constantly shifted beyond her subject as the boundary moved and Barad asserts both that matter is not discourse and the two are inextricably fused. Barad's 'cut' might also be interpreted as being a way of naming the moment when, rather than resolving discourse and matter into an observed phenomenon, the cut instead produces matter from the discursive by performatively making it into matter. In such a reinterpretation of the cut performatives of matter are performatives that attempt to name the unnameable in order to indicate that which is beyond culture and which cannot be part of social relations. However, this also changes the idea of the cut from something that resolves discourse and matter to being something that does not begin from or recognize such a distinction. A performative of matter can then be understood as the moment when a performative puts something into matter and beyond discourse by creating that very distinction. Such performatives of matter, instead of cuts or as reinterpreted cuts, begin from the discursive and look to understand how something that is taken to be non-discursive can be enacted.

In this interpretation, matter is made by a particular kind of performative through which we are able to talk about and act on that which is beyond social or cultural relations. The moment of mattering, of making matter, does not resolve a distinction of matter and discourse but instead creates a specific form for such a distinction. This means a performative of matter is the process of discoursing about things which we can know nothing social or cultural about. But this appears to be an impossibility because the moment discourse is capable of naming or acting on something that was beyond the discursive then by definition discourse brings it within the discursive and within language.

The performatives of matter thus cannot be pre-social and pre-discursive, yet they seem to be reaching for that very outside. Does this simply mean the performatives of matter are a constant failure? That they are the endless of task of pursuing an impossibility? This seems unlikely given the extraordinary power and depth of knowledge that, for example, scientific knowledges, which are the primary forms of discourse devoted to matter, are capable of producing and continue to produce. Matter as an impossible discursive pursuit seems unlikely given the success of science in pursuing matter and bringing it into social use.

Rather the performative of matter can be seen to be the means by which some norms, characteristics or practices are taken to be beyond the social and cultural. Matter is thus not a given, nor is it in contrast to discourse, it is instead the result of kinds of performatives that mediate when something is beyond the social, thereby creating in a specific context a discursive distinction between matter and discourse. It is a social logic to produce asocial things. These will be sedimented, well-known performatives, that produce discursive results of great power, for something beyond the social has to be taken as a given and remain unquestioned socially, simply because it is outside any social relationship. The great power of the performatives of matter will be to construct things that we not only can rely on but which we must submit to.

We can see this, for example, in feminist theory which, similar to anti-racist work, has often brought what were thought to be beyond social characteristics of the abject into the social, thereby making something that was matter, social again. And this is done because when it is 'matter' or natural that women are nurturing homebodies then there is no argument over whether they should work outside the home, it is simply beyond argument. This understanding of matter has similarities with that of the 'black box' in actor-network theory. In some such theories, it is argued that social relations are built out of a series of black boxes that close down and contain various social arrangements of actants (both human and non-human). All is built on series of these boxes that are held closed so that the whole arrangement can work or are opened up to disturb a particular arrangement (Callon and Latour, 1981). However, theorizing matter as a performative is more specific because for Callon and Latour all kinds of social arrangements, as well as material, can be black boxed. The account I am suggesting that makes matter a particular kind of performative is similar to black boxes, while being more specific in its definition of the cut-like moment when a performative of matter is performed and the specificity of the performative of matter in making something unquestionable and hence imbuing it with a certain kind of power.

The performative of matter consists of sets of practices that are repeated and cited by each other, in whose performance an object of the practice is put beyond sociality, thereby gaining a specific form of authority, or is brought into the social, thereby gaining a partiality and a politics. This is performative because the object of the citations is a materiality that is itself produced through the citations and which then becomes the origin of the citations.

This also finally clarifies what I mean by the social logic of performativity which always remains within the discursive. What counts as a difference and what counts as a repetition will be defined through such social logics, even while the abstract problem of difference that Derrida defines remains. Within performatives, both of matter and other kinds of performatives, we socially and culturally learn what may count as a repetition or as a difference, what is same and what is other. In this sense, nothing is outside the relation of care and capture and on this basis certain types of caring and capture, of play, resolve socially and make forms of being-with/becoming-with.

With this in mind, we can turn back to communication and try to theorize it fully, noting its three main conceptual foundations of presence, performativity and materiality.

A theory of communicative practices

Communicative practices are material performatives that create presence. Materiality is itself a particular kind of performative, one that does the work of splitting the discursive from an asocially authoritative sense of matter. This reflects the sedimented nature of performatives while also specifying which particular kinds of performatives, with their associated particular kinds of repeated citations or actions, such as signing a letter, constitute communication. Performatives are the sets of citations that create in their performance both the subject and object of the performance. The performatives construct presence in the interrelations between communicants that creates the communicants and their types of messages resulting in the possibility of a successful transmission of meaning.

Such communicative practices exist in the everyday, as part of repeated individual actions of communicants, while at the same time their repetition/iteration ensures they are part of historically specific and historically sedimented sets of practices. Communicative practices may be specific and local, perhaps expressed among very few people in a local slang or dialect, or may be global and generalized, perhaps expressed among, for example, the (in 2010) 12 million players of World of Warcraft. The abstract theory is applicable to a range of communicative contexts, meaning that it will be possible to identify communicative practices in specific places and times, operating with the social logic Butler identifies, each with their own sense of presence based on material performatives and creating the possibility of a particular form of transmission. This is the answer as well to Derrida's problem of repetition, because though repetition remains philosophically a problem each communicative practice will have to resolve this problem socially by identifying particular repetitions as significant enough to construct a socially stable form of repetition. To return to the question of pre-internet and internet-dependent communicative practices, now dragging along an extended theoretical framework, I return to the hypothesis and restate it.

Pre-internet communicative practices construct transmission based on performatives that put the body beyond social construction by employing performatives that identify an individual's subjectivity with their physical form. Presence in these communicative practices relies on and elicits itself through performatives that allow the identification of physicality with subjectivity, thereby stabilizing the communicants, the message and forms of messaging. The selves and others of this form of communicative presence come into being through the identification of a body with a subjectivity. Each performative and its everyday instances are subject to the necessary difference of repetition and iteration; for example, no two signatures are exactly alike yet their repetition is essential. This difference through sameness is held together in pre-internet communicative practices by social conventions and collective practices established in technological and cultural mechanisms such as the signature, forms of greeting and farewell, telling of personal stories, sealing with wax and so on. These are the core performatives that establish presence and so form transmission in pre-internet communication.

Communicative practices based on internet technologies construct transmission with performatives that identify the style of someone's transmissions with a subjectivity. Presence in these communicative practices is based on performatives that create identifications between particular types of messages and individual subjectivities. Each such identification is constructed out of performatives that attempt to move the question of which subjectivity to attach to which style of transmitting messages into a matter performative and so out of the social. Various social conventions and practices are created to stabilize transmission utilizing collective histories of types of transmitting of messages, repertoires of jargon, collective memories of individual types of transmission and so on.

Prior to the internet, communicative practices for at-a-distance communication attempted to replicate a physical face-to-face form of communication by assuming that the body and subjectivity were necessarily identified. Such an assumption needed to be put into practice by creating performatives that authorized the body/subjectivity of senders and receivers and hence the integrity of the message, whose meaning does not affect the body/subjectivity but is separated from them. Since the rise of internet-technology-mediated communication, communicative practices have developed in the face of a radical instability of markers that would identify a body and subjectivity. Instead practices have developed that stabilize sender and receiver by identifying each communicative subject with a style of communication. These two hypothesized communicative practices will now be focused on in turn, first with a study of letters to and from colonial Australia 1835–58 and second with a case study of communication in game-based virtual persistent worlds.

Notes

1 Derrida writes, 'One must therefore go by way of the question of being, as it is directed by Heidegger and by him alone at and beyond onto-theology, in order to reach the rigorous thought of that strange nondifference and in order to determine it correctly' (Derrida, 1976, p. 23).

2 Of course, this does not deny the importance and extent of Heidegger's work or comment on it. Nor does it deny the importance of the controversy over Heidegger's membership of and support for the Nazi Party (Derrida, 1989; Ott, 1993). The focus here however is not on overviews of thinkers' work but on taking up core concepts in the context of understanding communicative practices.

3 See Schatzki for a discussion of Dreyfus and Olafson, which is close to this point (Schatzki, 2005, pp. 239–41).

3

Letters: Pre-Internet Communicative Practices

I have been looking over all the letters that I have received to day, and there are only three in which the writer does not say that he (or she) has nothing to say, which is a most preposterous fact.

Henry Howard Meyrick

The letters I receive from home create much too powerful an excitement in my mind to call it a pleasurable sensation. It is a painful sensation occasioned by pleasure and memorie's softening influences over friendship, distant but not forgotten, that makes me deeply sensible of the strength of that tye that binds me to home, perhaps more powerful in its influence than even the allurement of gain or riches.

Niel Black

Introduction[1]

To be a colonizer in Australia in its early settler years was to be truly at a distance from family, friends and business partners who were more often than not located in Great Britain. Communication was reliant almost entirely on letters, with the only occasional adjunct being verbal reports when, for example, someone returned to Great Britain and was able to pass on first-hand news and reports of those still in the colony.

Letters from the period 1788 until 1872 (when Australia was linked to the United Kingdom by telegraph) were not only physically at a distance but they were also temporally distant, emphasizing that they travelled between two faces that were no longer visible to each other. One of the series of letters that will be discussed below was written by settler Henry Howard Meyrick to his family in Great Britain. Meyrick arrived in Melbourne, Australia, in 1840 and wrote 28 surviving letters to England (Meyrick, 1939). The fastest time a letter took to reach the post office near his family was just over four-and-a-half months, the longest was nearly eight months and the average was just over seven months. What could happen to a son and brother in the months between writing and reading? As we will see, Meyrick's letters reflect physical, emotional and economic changes of profound kinds, all of which had to be communicated through the medium of a letter. Letters to Australia from this period provide a particularly intense example of communicative practices in action.

The hypothesis of pre-internet communicative practices directs a focus on certain aspects of these letters. Unlike models of communication that exclude the content of messages, the analysis that follows includes such meanings. However, unlike many analyses of epistolary practices, the aim is not to understand those who are communicating and the content of their letters seen in relation to their intellectual, political or artistic achievements – for example, in the extensive correspondence between Mary Mitford and Elizabeth Barrett Browning (Milne, 2010, pp. 51–71) – but to understand how such communication is possible in the first place. This requires a shift in focus from relating the content of letters to the lives and contexts of the readers and writers to the forms of letter writing and reading, including their technologies and cultures. This does not exclude the contexts provided by the lives of readers and writers but it diminishes their importance, pushing forward comparison of how communication happened.

To achieve this, there will first be a quantitative analysis of what was found and what seemed to be key means of creating stable communication. Following this, three case studies from within the body of letters will be explored. One will be of the extraordinary record found in the letters of Henry Howard Meyrick. The second will be of the business letters of Niel Black. Finally, communication that came from the founding settlers of Melbourne will provide a view of letter writing outside of stable postal institutions.

Letters to and from Australia: 1835–59

Two hundred and six letters and seven closely related documents from this period were examined. The letters were of all types, from short arrangements to call, to matter-of-fact business letters to the most soulful and distressing accounts of death in the bush and of genocide. Such a mix is appropriate to examining communicative practices as it offers a range of types of communication, hoping to find what is common between them. To keep a broad track of the differences between types of letters, they were designated in four categories; personal, business, personal and business or state.

Each letter was read, with nearly all letters being handled in the original, this was an important consideration given some of the material aspects of letters that needed attention, and letters were coded along two axes. One axis attempted to distinguish the particular materialities and technologies used in the letters. Materialities and technologies were interpreted broadly to include not only types of paper and methods of folding but also the facts of signatures, and the placing of greetings and farewells. The second axis attempted to identify various routines within the meaning of letters. Here analysis tracked such things as whether self-identifying anecdotes or recollections of personal history or tales of personal exposure were present. While interlinked, the difference between the two axes is exemplified by the fact that in both axes greetings and farewells are examined; in the first axis this is simply the fact of there being in a particular place on each letter a greeting and farewell, whereas the second axis traces the content and nature of farewells and greetings (Table 3.1).

In terms of institutional context, this period is one in which at the beginning the postal service charged per piece of paper sent and not in terms of weight.

Table 3.1 Coding categories

Materialities	Meaning routines
1 inks	1 position/form greeting/farewell
2 writing implements	2 self-identifying anecdote/personal history
3 form of envelope/letter folding	3 track personal history across letters
4 position/form greeting farewell	4 levels of self-exposure
5 seals and signatures	

The most common letter was one sheet of paper folded in such a way as to write the address on the exposed section while the message was written, usually, hidden inside the folds which were closed by a wax seal. Later in the period, letters which were contained within envelopes emerged. This occurred variably; for example, within one sequence of business letters from 1859, letters sent from Melbourne continued to be folded and sealed single sheets of paper, while from the same person at roughly the same time but now in Sydney letters appear to have been placed in envelopes. Within one sequence of letters sent by the Bank of Australasia on 19 May 1859, the letter was sent using the single, folded piece of paper technique, and on 20 May, envelopes began to be used and continued to be so (MS8996 Folder B-Ban Black letters). Coding of materialities leads to a number of quantitative measures that are worth comment (see Table 3.2).

Though rounded up to 100 per cent, there were in fact three exceptions to a greeting and farewell placed in a letter, though these exceptions demonstrate the rule. One instance occurred in a letter that was contained within another letter. This dealt with the death of a daughter, whose mother was sending a letter trying to find her son; the letter is distraught and not signed by the mother. This mother's plea was contained within another letter sent by the missing son's uncle, which explained the situation and which contained a greeting and farewell (MS8996 Box 20, 17 September1849). The other two instances were invitations from the aristocracy to visit, the inviting Duke did not put a greeting and though there appears to be a salutation at the end there was no signature, and it is possible these were hand delivered (MS8996 Box 20, 14 June 1843; 13 June 1848). These examples underline how unusual a lack of greeting

Table 3.2 Materialities: percentage of letters with material practice

	Folded	Greeting at start and farewell at end	Signature	Seal
Total	71	100	99	71
Personal	91	100	98	91
Business	59	100	100	58
Mixed Personal and Business	80	100	96	80

and farewell was in framing the content of a message. The greeting and farewell in their near universality stabilize the content of the letter by indicating where meaning starts and stops. Even the examples of postscripts confirm this as they are strongly indicated by a smaller greeting (often a dash or a 'ps') and are often themselves closed by a shorter version of a farewell (sometimes just initials). The content or meaning of the letter is in this way bounded and identified.

Seals and signatures establish an identity for the author of the letter by the physical marking of the letter with ink and wax in a way that has no other meaning than to refer to an identity. In these practices, the inevitable uncertainty about who is writing at such distances is contained by the assurance that this particular body – of a brother, son, friend – touched this letter and accordingly is the author of its contents, themselves bounded by a greeting and farewell that further marks out the identity of the author. These technologies seem to strongly resolve several key uncertainties, thereby providing some basis for the emergence of presence between writer and reader(s) and so for a transmission in which the content is understood to retain its self-identity and come from the body of the author to the reader(s). The lower percentage of letters with seals mirrors the distribution between numbers of folded letters and letters in envelopes. No envelopes were found during the research and accordingly it is impossible to tell if seals were used on envelopes or not. When handling folded letters it is easy to verify if a seal was used due to markings on the paper, even if the seal has been lost.

In contrast to these universal and near-universal techniques, some other material techniques were not consistently found. Issues related to ink and paper were tracked but not in relation to whether they were used or not, because this is obviously rather a self-fulfilling criteria (no exceptions were found to the use of some form of ink and paper). However shifts and changes in their use or mentions of difficulty with them were tracked with the result that there were very few mentions of ink, pens or paper. A rare mention is by Meyrick, who at one point states, 'I am far from sure that you will be able to read a word of this, as I have manufactured the ink out of blacking and have got a steel pen which runs thro' the paper at ever third word' (MS 7959 H15789–H15816, 20 March 1841), though the letter was, even nearly two hundred years later, entirely legible. Meyrick also mentions using a black swan's quill though he considers 'The black swan's quill does not succeed' (MS 7959 H15789–H15816, 10 September 1841). Nowhere else were there found mentions of the basic technological items of ink, paper, sealing wax and so on.

Similarly, potential changes in handwriting were tracked but none were obviously found and no mention was made in any letter of someone's handwriting, whether of it improving, worsening, staying the same or anything related. It was however noticeable while handling letters that, and this may seem obvious but it was striking, the signature and handwriting were clearly thematically linked being born from the same style of writing of an individual. Thus the handwriting, whose mere form must have been familiar to some readers, was reinforced and confirmed as identifying someone through the 'look and feel' of the signature being close to that of the message. Rather than presenting this as separate quantitative result it is integrated into the signature's role in establishing the body that writes, because of the role of the shape of handwriting in reinforcing the signature's relation to the handwriting in the message.

This analysis establishes an initial uniform set of practices for writing letters. These are a greeting and a farewell, placed at beginning and end, a signature on all and a seal on folded letters, with contributory practices of identification between the shape of handwriting and the shape of the signature. There were other practices we might intuitively have expected for which no routine or pattern was found. Two such are a date and location placed at the start of the letter above the greeting. Dates were irregularly used in letters, both in terms of actually stating the date and where it was stated. For example, dates might be mentioned as part of the letter's discussion of an event or in identifying a business deal but it was usually unclear if this was also the date of the letter or near it, so it was unclear if dates were part of the meaning of letters or of what I am separating out here as material practices. All that was clear was that a practice of writing a date near the greeting at the beginning of the letter was not routine. It may simply be that writers and readers were able to assume the date of the letter from the postmark from when it is sent, after all with folded letters the postmark would always be present, unlike with envelopes when the letter might be retained and the envelope with the postmark discarded.

Similarly, stating where the letter was written from was not a routine practice. Again, letters variably stated locations, though with no clear pattern of stating where the writer's location was, but this appeared to be more closely related to the telling of stories within letters than to a routinized practice of stating the location. Again, postmarks usually stated locations and it may be that no need to state the location was felt as the postmark would be on the same piece of paper as the message.

Even more so than the previous category of material practices, practices of 'meaning' as communicative practices are primarily a qualitative set of procedures and routines. For example, it was not possible to quantitatively code the shifts in the content of greetings and farewells as these were closely dependent on their contexts for their significance. Accordingly, more will be said about these kinds of practices in subsequent sections. However, three kinds of broad routines of meaning were able to be coded (Table 3.3).

First, it was possible to identify in a general way when the content of letters contained self-identifiers. These were remarks of the kind that could only be made by the writer of the letter, and so performed a function of reinforcing the identity of the sender. In business letters, this often took the form of mentioning other occasions when the writer and receiver had made business deals with each other. Many of the combined personal and business letters became personal because the writer drew on some personal connection or knowledge to establish their identity, and sometimes to try and persuade. The difference in percentages between business letters and mixed personal and business letters for this category is therefore something of an artefact of coding, because mentioning some personal self-identification anecdote or story was

Table 3.3 Cultures: percentage of letters with routines of meaning

	Self-identification	**Reference to other letters**	**Personal self-exposure**
Total	63	59	60
Personal	98	72	96
Business	38	46	34
Mixed Personal and Business	92	88	92

one of the key ways of transforming a business into a business and personal letter. Yet, even so, nearly 40 per cent of purely business letters contained some kind of confirmation of personal identity.

Second, writers often referred in one letter to other letters that they had sent or had received. There was often an anxiety about maintaining contact and understanding the sequences of letters. This is perhaps unsurprising given the distance letters travelled and the time they took – as noted already, the Meyrick letters range from just under five to nearly eight months – and the fact that those moving around the colony, either for sightseeing or more likely as they try to find a better position to make a new economic life, may not pick up their letters for some time. One way of establishing communication is to talk about the specific communications that are going on between individuals and groups. It was therefore possible to code letters that made explicit reference to other letters sent and received.

Finally, the most difficult coding exercise undertaken was to identify in letters when writers made clear displays of their personal identity. This is not the use of anecdotes or stories that were tracked in the first category, but the explicit self-exposure of states of emotion, of bodily changes or developments or of ways of expressing the writer's understanding of their life and changes they are coping with. In contrast to telling stories that both writer and reader will recognize as bits of memory that will bind writer and reader together, here it was noticeable on reading letters that many writers were reaching out through the ink by discussing their lives. This is to be expected as, after all, what is a letter for but to keep one's identity in front of the identities of others. Yet, what was looked for here was not the simple recounting of any event, such as a reflection on the weather, on government regulations, current economic conditions and so on, but clear and explicit reflections on a bodily or emotional state of the writer. Despite the obvious difficulty of identifying these clearly, it was such a striking practice, even showing up in a third of business letters, that it seemed worth trying to quantify. The overwhelming numbers of mixed personal and business and personal letters reflect the importance of this practice.

Keeping in mind the difficulty in coding quantitatively some of the phenomena touched on in the tables above, it is still striking that a number of practices can be considered universal across letters. These practices, in their universality, form something like an unthought binding for messages, establishing some of the basic practices that all those involved in these kinds of communicative practices used so commonly as to render them obvious.

This binding consists in a number of elements that establish both the placement of messages and the identity of sender and intended recipient. In terms of stabilizing where the message can be expected to be found, the placement of greeting and farewell are key. Further the greeting establishes who the intended recipient is and accordingly provides clues as to how the writer is expecting the letter to be read. We will see in examples in following case studies how writers shift their form of address and content depending on who the intended recipient is. Similarly, the farewell is universal and provides an identification of the writer, particularly when linked to the signature, and closing the contents. This is marked in some letters which have either a lack or an abundance of things to write about. In the former case, there are blank spaces in the letter making the farewell and signature necessary to tell the reader that the message has ended. In the latter, the normal practice with folded and sealed letters was that only one piece of paper was used but this did not prevent the writer writing on every piece of unexposed paper and then sometimes turning the letter perpendicular and writing over the top of existing writing creating a cross-hatched effect. In these latter cases, the writing can become so confusing that a farewell and signature is essential to identifying where the message finishes and which lines need to be distinguished as having been written at which angle.

Reinforcing these binding and universal practices are a series of other practices which particularly for personal letters ensure the person who is writing is identifying themselves to the readers. While it may seem obvious that individuals will refer to themselves and expose their inner feelings in personal letters, it should not be overlooked that these are done in a context which identifies one person as the writer and which imagines one or more people as the readers. Similarly referring to previous letters, to letters just received, complaining of not receiving letters or warning that few letters will be forthcoming constructs in-between the letters a connection that imagines an ongoing relationship between the writer and his/her readers. Why business letters require less of this most likely relates both to keeping matters away from the personal, and in this sense is partly an artefact of the construction of coding categories, while also gesturing towards business as dispassionate and opening the question of whether business letters reinforce in some way the identifications of writer and reader to compensate for some of the content which is found in personal letters. This latter question will be taken up in the largely business letters of Niel Black.

These routines provide a series of strong binding practices which locate the content of a message as emanating from one body or subjectivity and that the intentions of this body imbue who the letter is directed to, and hence how it is written and to be read. The communicative practice suggests that presence will be created and stabilized based on these material performances, the strong bindings, which create a body and identities for the writer and readers, on the basis of which transmission is made possible. One example is the constant presence of greeting and farewell which frames the message that is to be transmitted, clearly establishing where the writer believes the message starts and finishes and indicating this to the readers. With this practice in place, the message is in part created and meaning may be transmitted. The strong bindings are powerful in their universality and their characteristics are to construct a writer whose fingers have guided the pen's nib and traced the ink on the page; they confirm the body touched the paper that conveys the identity. In this way, a sense of presence between writer and readers can be created which stabilizes communication and produces the moment of transmission, when a reader can take meaning from the page as conveyed by the writer. At the distance of months in time, thousands of miles in space and gulfs of culture between the centuries-old United Kingdom and the years- and decades-old colonial Australia, we find a writer identified as having touched to produce a message whose readers are able to comprehend instantly through stable and binding communicative practices.

While recognizing that what are fundamentally qualitative routines have been quantified and that this must inevitably be somewhat crude, it is striking that various cultures of meaning repeatedly occur in letters in ways that stabilize the identity of a writer and attempt to connect this to the identity or identities of readers. If we take these cultural practices and place them together, as they are in every letter, with the more mute material practices, we already gain a view of communication in which writers and readers are familiar with a range of routines and procedures that can produce simple markers of identity alongside complex enunciations of identity. To begin to draw on this quantitative introduction, it will be useful to see letters as communication in action in three case studies, starting with the remarkable series of letters written by Henry Howard Meyrick from colonial Victoria to his brother, sister and mother in the United Kingdom.

Henry Howard Meyrick: 1823–47

H. H. Meyrick[2] left a series of 24 letters tracing his arrival in Australia in 1840, when he was 17, travelling with one cousin (Alfred) to meet another cousin (Maurice) who was already in Australia and carrying £1,000 in letters of credit. During this time, his letters trace several attempts to set up his own farms, including failure when being swindled out of one 'run' he thought he had bought, having to take a loan from his UK family and finally looking like he was working on a good prospect. He was initially helped by indigenous people who showed him how to make shelters and to eat kangaroos on his first visits to wild land, through to his provision of a clear and unambiguous account of a core controversy in Australian history in the genocide of indigenous peoples as settlers tried to claim land for farms. He died seven years after arriving when pressed to cross a river in full flood on horseback to fetch a doctor for a friend whose wife was in labour; he was swept away and drowned. Horribly and ironically, he had already reported on a similar death only a year before his own, when he wrote of 'Stewart, he was about to cross a swollen river, and to those who tried to dissuade him he answered "I will cross tho I should go to hell on the road" he pushed his horse in, and drowned' (2 August 1846), which with different motivations was what happened to Meyrick (Meyrick, 1939, pp. 228–36).

The following explores the way the strong bindings identified in the previous section played out in Meyrick's letters. For someone like Meyrick there are a range of practices that are universal and that establish the identity of the marks on the page to his body. However, this universality still has to be taught; Meyrick remarks in his third letter that 'Ann will perceive that I have profited by her lesson in the art of folding letters' (20 June 1840). Keeping this strong binding in mind, there are then a number of ways of traversing Meyrick's letters to draw out the communicative practices he is utilizing; there are the forms of greeting and farewell and how they shift; there is the anxiety about letter delivery and keeping sequences; there is the maintenance of the story of the writer; references to mutual acquaintances and there are moments of acute emotions. Across these it should be assumed that the strong bindings are present, as it would be tedious to constantly refer to them. We can now reach into some of the greater complexities that Meyrick's skill as a writer and the survival of a full series of his letters to his family, though not unfortunately of replies to him, allows.

Early greetings and farewells in letters of Meyrick follow a pattern. To his mother he writes to 'My dear mama' and to his brother 'My dear Jim'; in both cases, he closes with 'your affectionate brother' or 'your affectionate son' and signs HHMeyrick. The seventh letter of the sequence to his mother shifts to a simply signed 'Henry,' which remains consistent for several letters before reverting to the full signature of HHMeyrick, though at the same time he closes a letter to his sister 'your affectionate brother, Henry' without the full signature. Even later, he begins to simply initial letters to his brother Jim with a closing 'your affectionate brother HHM'. He also returns to using 'Henry' in some letters to his mother. Can much be made of this? Perhaps only that the room for manoeuvre over his name and signature within these communicative practices is highly limited. Were he to sign at any time outside of practices that identify him immediately – for example, is it inconceivable he might have playfully signed 'Lord of the Bush' or some similar? – then the management of any such uncertainty would have to traverse months and miles. Even so, we can imagine that the strength of other bounding routines, like handwriting and conveying news, would be strong enough to cope with some playfulness around the signature, perhaps it was just not Meyrick's way.

A further constant presence in Meyrick's letters is the arrival and non-arrival of letters. He effectively keeps a tally of how often he has written and received and requests confirmation that his letters have been received and of how many have been sent to him. In one exchange, he first complains that 'I think myself a considerably dutiful boy, having written a dozen letters, and only received three' (1 November 1840), but only ten days later, he is writing a letter to his mother in which he lists a number of new letters he has received asking for confirmation if he has missed any and noting how satisfied he is with the influx 'as it was but a week ago that I was blowing you up for not writing' (11 November 1840). It is a constant of his letters that he mentions what he has received and confirms what he has sent, and keeps them informed if 'I have not written to you for some time inasmuch as I have not lately had the means so to do' (16 May 1841). The anxiety seems to be reciprocated as Meyrick's letters imply that those in England worry over his health and well-being, as is to be expected for many reasons, not least because there are a number of stories of other people's deaths in Meyrick's letter. At one point he replies to his mother:

> I cannot understand how it is that you have received no letters for so long a time, as I have always written regularly, & I do most earnestly beg and beseech you if such a thing should happen again, that you will not fill your mind with such death's head and cross bone visions, for be assured that we shall never take any important step of any sort or kind without first acquainting you of our intention, and if any evil should befall us, we should never think of concealing it for one moment. (18 October 1845)

In Meyrick, it is a routine to ensure communication is asserted in the interstices between letters by constant reference to their travel and arrival, or not. The effect of knowing what has been received and sent is to reinforce and reassert that these subjectivities are who they are. This is closely related to the further practice of updating his story. This is another of those points that seems so obvious that it is a wonder it is worth paying attention to; of course, a son writing from the colonies will tell what he is doing and will do so sequentially leading to a story that is updated. However, like the signature, that something is obvious does not mean it is not also significant, and here Meyrick's identity is constantly updated as he recounts his attempts to start a farm. One example of this, which surely would have been of particular significance to his family in England, is his relationship with his cousin Alfred who came out with Henry on the same boat and were met on arrival in Melbourne by Alfred's brother

Maurice. There is initially little mention of Alfred or Maurice until Henry outlines his current farming venture and notes:

> you will observe that there is no sort of partnership between Alfred and myself, and that if I like to spend my money upon grog it is nothing to him; we merely let our stock run together, and bear the experience of the station between us. (20 June 1840)

There then follows across his letters a series of notes about his cousins in Australia, which usually consist of their business relationships that seem to come together and move apart at different times. In this vein, there are notes of when he is in Melbourne picking up goods while his cousin keeps an eye on his run of sheep, or of times he is driving their sheep together or when he has not had a letter but he expects Alfred has them and he has not seen him for some time. Again the threads, as obvious as they may be, are of a familial identity asserted and established.

The difference in tone of Meyrick's letters also asserts a familial connection because he writes of some different occurrences to his brother and sister in relation to Maurice in Australia. For example, he mentions first to his brother Jim that Maurice has become engaged, which he does not mention to his mother till nearly a month later and there is no other mention to his mother until to his sister he notes:

> What do you think of Maurice and [indecipherable] she has behaved very ill in the transaction, and was making a fool of him all the time, in order to bring the other fellow up to [indecipherable] and then she dropped the country cousin. (15 January 1847)

While these are rather fleeting bits of evidence, it suggests a certain management of his audience when writing. Further evidence in this regard is his constantly upbeat assessment of his prospects to his mother. Even when recounting difficulties, Meyrick always notes how these do not apply to him, and that he has planned well and is doing well. For example, in two successive letters that are two months apart he mentions he has been ill but in both cases he is fine and well again, despite the second letter detailing the same illness as the first and referring to himself as having been reduced to a 'skeleton'

(07 January 1841; 20 March 1841). My reading of this upbeat self-presentation was done solely from reading Meyrick's letters; yet a 1939 book written by his nephew, F. J. Meyrick, based on both Meyrick's and some other letters not now available and on family history emphasizes that Meyrick was very sick during these times and had difficulty recovering, something I only realized after reading Meyrick's book despite having read the original letters (Meyrick, 1939, pp. 144–8). If this is the case, then the constantly upbeat tone of his having just beaten his illness contrasts strongly with a deeply difficult period of illness recounted by F. J. Meyrick and further suggests the management of self-presentation in the letters.

The last evidence of this management of identity is that, in another sequence of letters, he mentions difficult economic times in Melbourne, but in one letter states he will weather them fine as he has wheat and oats to sell and they will fetch a good price. In a letter some months after, he again mentions the ongoing economic difficulties, even suggesting that unless things change 'the kangaroo will again drink at the Yarra' (25 November 1841), while admitting that wheat and oats did not sell well but now noting he will be fine as he has potatoes (10 September 1841; 21 November 1841). Again F. J. Meyrick's account is one of considerable economic difficulty with the collapse of an economic bubble in Melbourne, leading to extremely difficult times generally (Meyrick, 1939, pp. 156–66).

Meyrick's general positivity to his mother, entirely understandable, breaks down on two occasions where he drops the more usual self-presentation of achievement in the face of difficulties for a direct emotional appeal. The intensity of these two occasions shows two things for the present argument; on the one hand, it demonstrates how the strong binding allows the real anguish and fear of extreme emotion to be conveyed and, on the other hand, we can see that strong binding is asserted even in a period of acute stress.

The first of two letters is an account of genocidal practices against indigenous people. Early in his correspondence, Meyrick refers to having been taught by aborigines to build miah miah huts in the bush and how to survive by living off the land (16 June 1840). Five years later, he refers, almost in passing, to how aboriginals are dying off, 'rum is killing those who we civilised, and the . . . men are killing those who are not' (18 August 1845). In neither case is this treated in a way different to the passing of other news such as stock prices, family news and so on. However, in 1846, Meyrick moved to Gippsland in his ongoing attempt to carve out a new farm from the wilderness and here he details genocidal practices, his helpless disgust and the pressure to help with them. Early on, the letter is simply another account of his development of a farm, this time emphasizing the difficulties of setting up. He then, quite abruptly, turns to the shooting of indigenous people. He notes the killing is being kept secret as killing women and children 'would lead to a hanging,' and that a settler friend who refused to kill defenceless indigenous people was subject to indignation and pressure. He goes on:

> For myself if I caught a black actually killing my sheep I would shoot him with as little remorse as I would a wild dog but no consideration on earth would induce me to ride into a camp and fire on them indiscriminately as is the custom here whenever the smoke is seen. They will very shortly be extinct it is impossible to say how many have been shot, but I am convinced that not less than 450 have been murdered altogether – I remember the time when my blood would have run cold at the bare mention of these things but now if I am become to familiarised with scenes of horror – from hearing murder made to a topic of every day conversation – I have heard tales told and some things I have seen that would form as dark a page as ever you read in the book of history but I thank god that I have never participated in them. If I could remedy these things I would speak loudly though it cost me all I am worth in the world but as I cannot I will keep aloof and know nothing, and say nothing. Remember me to Harry and Susana your affectionate son Harry. (20 April 1846)

The letters immediately prior to and after this letter (both only a few weeks apart) have no mention of these or similar events, which modern histories have gradually confirmed as all too common (MacKellar, 2008, p. 24 and pp. 200–2). The irruption into the normality of his series of letters recounting his attempts to build a new life is quite striking. Also striking is his normal beginning, which first covers his life and some small news of others before nearly two-thirds of the letter covers the genocide and his sudden end with the normal salutation and signature (quoted fully above). The intensity of his situation is in this way bounded by established communicative practices.

The second similar irruption is to do with the death of a friend and close worker, G. B. Eagle. This letter uses his normal greeting and farewell but is almost entirely given over to the account of Eagle's death and its consequences. Meyrick recounts Eagle feeling a little unwell and lying down for a rest. Sometime later Meyrick checked on him only to find him near death, dying 30 minutes later; 'Mother the near approach of death is awful at all times but in the silent forest, in the solemn stillness of the bush, its terrors are redoubled' (08 August 1846). Meyrick recounts what a good friend Eagle had been. He also notes he has to write to Eagle's sister to dispose of Eagle's property and he spends some time offering suggestions on how his mother might be able to find Eagle's family in England and send them news. Meyrick then dwells on the fact that he has been close to three sudden deaths recently recounting the already-mentioned death from crossing a swollen river and another sudden death.

Again, this is an emotional and unusual letter in his normal series and particularly unusual in conveying, like the letter about genocide, to his mother his difficulties and exposing physical danger and emotional turmoil. It is clear from letters at this time that he was having economic difficulties and the death of Eagle would create more as he lost a co-worker as well as a friend. Yet, there were many difficulties in Meyrick's letters, and apart from these two letters he presents himself as a subject in control of his destiny, which if a struggle was also a struggle moving towards a better life.

Meyrick is an interesting source both for the occasional eloquence of his letters (and for his readable handwriting) and for the ability to follow someone from their arrival in a distant colony and how this is played out in his letter-based communicative practices. There is enough detail and richness to Meyrick's letters to allow us to see how the universal practices can bind together his identity and project this across years and thousands of miles to his family. He occasionally mentions his body or appearance – once describing himself, once sending a daguerreotype home and claiming it is a poor likeness – but these are incidental to the practices that bind the body that touches the paper to the message in the paper to the readers so far away. Nothing symbolizes this stabilization of persona and body to message as when he requests three sets of Wellington boots, then quite fashionable, but mentions no size. His body remained just that, the body of H. H. Meyrick across all that distance and time, all the difficulties and adventures, until crossing a swollen river to fetch a doctor he was drowned.

Niel Black: 1804–80

If Meyrick represents the tragic settler then Niel Black is a different kind of story: the hardworking, thrifty success story. Not only did he live nearly three times as long as Meyrick, he married and had three children, became a Member of Parliament and on death in 1880 left an estate estimated at £179,208 (which would have been worth around £13 million in 2008) (MW, 2010). Black was a Scottish immigrant to Australia, arriving in 1839 in a partnership with two others, one of whom was a first cousin of the future British Prime Minister William Gladstone. Black surveyed pastoral prospects near Adelaide, Melbourne and Sydney, settling on Melbourne where in 1840 he bought a 17,000 hectare station. He added several times to this, remaining a success even when the partnership dissolved, and he received sole control of a third of the business. Naming his main estate Glenormiston, he was visited by Prince Alfred, second son of Queen Victoria, during a royal visit in 1867 and served as a member of the Upper House of the Victorian State Legislature (Kinross-Smith, 2002; MacKellar, 2008; Ward, 2010).

Black also left behind an extensive archive of letters and documents but, in something like an inversion of Meyrick, these are primarily of letters to Black rather than from him and primarily about business and not personal matters. This extensive correspondence provides a business counterpoint to Meyrick's personal collection, even taking into account the different direction of the letters. It is then worth exploring this archive to provide a counterpoint between personal and business communicative practices, allowing some attention to the places where the quantitative analysis indicates possible differences or similarities between such practices. Again the key work of binding practices will be important to tease out, but at the same time it will be useful to examine something of the quality of exchanges that largely put aside personal matters. Black also left behind a journal of his early years, which provides some context for the letters; however, as the subject here is the practices that make communication possible, this journal and MacKellar's commentary on it were, like Meyrick's nephew's book, only read subsequent to analysing the letters (MacKellar, 2008).

There are two contrasts between the Black business correspondence and Meyrick's personal that are worth noting at the outset. Business letters are in the main briefer and far more to the point than personal letters. This may seem obvious but it is not necessarily the result of focusing on a business interaction, as there are myriad details in a deal – often stock selling and buying for Black – that might be the subject of detailed letters. However, such complexities that might have been detailed are usually not and letters tend to stick to the primary qualities of a deal or topic. Second, what I have called binding practices are both universally present but are also generally more ritualized and less changed. Where Meyrick, and some other personal letters examined, do at times shift their signatures to more personal forms and sometimes add postscripts that may or may not be marked with initials, such things are not the case in business letters. Whether this reflects a greater nervousness in business letters meaning that with financial matters at stake the binding had to be reinforced and so standardized even more than in personal letters or whether this simply reflects an economy of practice is difficult to tell. For both these points, the emblematic letter is one to Black which has the normal greeting, farewell, signature, folding and seal, but whose contents in their entirety read: 'We are sorry to informe you that some of our sheep are scabby' (Black, 15 March 1855).

The greeting, farewell, signing, folding, seals and envelopes were uniform enough, and the content often arid enough, that in examining groups of letters there was little to note except adherence to the binding routines. Where there is a series of letters, the forms of greeting and farewell largely stay the same, with only minor variations. Very few letters include personal content, and this reader began, ever so slightly, to become tired by the constant reference to the price of cattle. Yet all this might be as expected from business rather than personal letters, but it should be emphasized that these business practices are subject, just as personal letters were, to time delays and huge distances in space. As the difference between Black and Meyrick's careers demonstrates, there were not only potentially large successes to be made but also long grinding failures, injuries and death. Some practices also suggest that there was even more emphasis in business on maintaining clear binding routines.

A sequence of letters in 1850 that are preserved from Black's then-partner T. S. Gladstone in Liverpool suggest that anxiety about communication led to some stronger practices than seen in personal correspondence. The unusual practice is that each letter contains a copy of the previous letter that was sent to Black. Letters are folded and sealed according to pre-envelope practices, meaning that the sending of two pieces of paper in one would have incurred double the cost of postage. The letters, both the originals and doubles, retain full binding practices, with signatures on all (including the copy) and the same greeting and farewell: a greeting of N. Black esq, Dear Sir, and a farewell of 'your obedient servant'. Part of the way through the sequence the conversion to envelopes seems to occur as multiple pages of unfolded and unsealed letters appear (Black, MS8996 Box 17, Folder 1).

Further, the copied letters are clearly written in a different handwriting to the original letters. The handwriting is smaller and formed more neatly, making the copied letters easier to read. Further, the copied letter often seems to suffer from some 'spreading' of the ink, making each individual word appear a little smudged, though it is impossible to know whether this is how the original would have appeared to Black as he opened and read them for the first time. What this smudging however does confirm is that different inks and possibly paper are in use as well as a different hand. Further, the copies are signed with the same signature as the originals, suggesting that Gladstone used the signature to authorize to Black that these were true copies. This might seem a detail but remembering that Black should have already received the original, Gladstone who then sends the copy would never know if the copy and original or copy or just the original would arrive and accordingly would never know that any divergence between copy and original could be checked or not. The signature thus seems superfluous as faking any part of the copy ran the obvious risk that the original would already be in Black's hands. No matter though, for the copy is signed (Black, MS8996 Box 17, Folder 1). This example from amid the correspondence to Black from, at that time, his most important business correspondent suggests that even the smallest potential for loosening the routines that bound message to sender to receiver was of concern and was answered by strengthening the same routines.

A further window that Black's archive provides is into the hybrid category of combined business and personal letters. There is a 23-letter series to Black from Donald Black, a cousin, in which the personal and business are clearly mixed at first, reverting after a dispute to standard, terse business practice and then gradually returning to a mixture. The sequence begins in 1851 with Donald Black clearly having some familiarity with Niel Black, opening his letters (and also closing them) with the greeting 'My Dear Mr Arch'; the slightly ungainly 'Mr Arch' perhaps reflecting a personal familiarity based on family mixing uneasily with a business setting, as the letter mainly concerns the business of having a land survey done (Black, MS8996 Box 3, Folder Black).

The correspondence continues primarily on matters of business with occasional interjections from Donald Black. The familiarity seemingly growing as 'My Dear Mr Arch' moves to 'My Dear Arch'. An example of such personal interjections are nearly a year into this sequence of letters, Donald noting, 'I am in receipt of your letter on the 7th instance and in reply to your letter I have wrote to you ascertaining the melancholy accidents of Mr Ewen and also of your narrow escape of the same fate' (Black, MS8996 Box 3, Folder Black, 17 January 1852). This letter also states that Donald has paid his debt finally and offers Niel some almost-parental advice, 'You ought to have a little more patience in doing business of that sort between you and your uncle. You are yet but a young man and whatever his advice to you, you may depend that it is for your good' (Black, MS8996 Box 3, Folder Black, 17 January 1852). While in no way comprising the majority of the content of these letters such interjections are present until a sequence of two letters. In the first, Donald registers his surprise about a deal Niel has done to sell some of his cattle that involved a delayed payment to Donald and not cash on delivery. While Niel's response could not be found, the next letter from Donald quickly backtracks emphasizing that he was simply asking when he was going to be paid, and he was not criticizing the deal. From comparing the two letters, the later letter appears to be a fudge to cover over his criticism in the first letter in which he certainly welcomes the deal but is concerned about not being paid immediately (Black, MS8996 Box 3, Folder Black, 22 October 1853; 18 February 1854). As if to emphasize the time delays in such matters, these letters, which even today can be read as urgently referring from the second to the first, were four months apart in writing.

At this point, there is a return to 'My Dear Mr Arch' and letters take on a more business-like tone. From this point on, there is certainly no repeat of the familial kind of advice, or the use of a post-script that is initialled dealing with a personal relation. The matter of the cattle being sold continues, such that Donald creates a nickname for them as 'the fat cattle'. However, from the more familiar and even parental style, Donald shifted to less familiarity, accompanying a number of times when he seems to be in the role of supplicant, for example when introducing his nephew or when trying to arrange to meet Black (Black, MS8996 Box 3, Folder Black, 08 August 1854; 06 October 1854; 12 January 1854).

It is tempting to read something of the shifting relationship between Niel, who in this period is clearly becoming a rich and substantial success, and Donald, who unfortunately I could find no context for other than his letters. Similarly, it is tempting to read the letters as a sequence in which Donald begins as a familiar but then particularly through the dispute over cattle, he moves to a more formal footing. Such an interpretation is buttressed by Mackellar and Kiddle's accounts of Niel Black as a formidable individual, but it is difficult to fully justify such an interpretation on the letters I examined (Mackellar, 2008; Kiddle, 1961).

What remains consistent through such potential shifts is the use of all the communicative practices so far identified. Moreover, these practices remain consistent across the widely divergent prospects and subject matter of two like Meyrick and Black's correspondents, as well as coping with the temporal and spatial pressures of colonial correspondence prior to the telegraph.

The Port Phillip Association

To complete this review of communicative practices in colonial letters, one further set of documents was examined in order to provide a contrast to the majority of letters examined. This was the set of original documents from the

Port Phillip Association that refer to the founding of Melbourne. This story has received extensive historical attention, and it is not the present intention to add to the historical story. Instead, the set of documents covers a time and space when there was no institutional framework for communication. It accordingly offers the possibility of seeing how communicative practices in 1835–6 operated outside of postal institutions. This choice was partly made for practical reasons, as the documents were in the same library, State Library of Victoria, Melbourne, as the Meyrick, Black and other letters that have been consulted. Unfortunately, this also brings a limitation in that this is not an extensive archive. There are only 14 documents and half of these are minutes, copies of a journal, a list of travellers and so on that are not amenable to analysis alongside the letters that this chapter has so far dealt with. However, the seven letters that remain provide some insight into a fast-moving and deinstitutionalized situation in which communicative practices were still vital.

First, it will be useful to briefly recap the historical context. The Port Phillip Association was a commercial association that sought to conclude a treaty with indigenous peoples to take over a significant amount of land in what is now Melbourne, Australia. This has some significance as it is one of the few, if not the only, occasion when negotiation with indigenous peoples was undertaken during British settlement of Australia (unlike negotiations with indigenous peoples in New Zealand). John Batman claimed to have negotiated a treaty that would make him a rich man, leading to a second company mainly associated with John Pascoe Fawkner being formed that entered into competition with the Port Phillip Association. As the British government feared this contractual relationship with indigenous peoples might undermine the basis for property rights in already-colonized New South Wales, the Port Phillip Association's contract was declared null and void, and the settlement illegal. However, while government officers communicating between London and Sydney tried to develop a plan to deal with the settlement, further stock and settlers arrived. In the end, this led to the legitimization of the settlement as a British colony in 1836 (Macintyre, 2004, pp. 66–78). The documents surveyed relate solely to the first two years and primarily to the attempt to either legitimize or delegitimize the initial settlement.

Across the seven letters surveyed, it is clear that an institutional context for mailing letters was absent. Less than half are folded, and only two of the seven are sealed. Rather than a consistent practice, it appears that because letters had to be hand delivered by those travelling to and from the new settlement, sometimes they were folded and sealed, though not on regular-sized paper as is seen in letters from later times, and sometimes they appear to have simply been folded and given to a messenger. Unfortunately, there is no evidence for or against the use of envelopes, except to mention it would be close to 50 years ahead of their first use found in this research.

The consistent practices across all letters here is that of signing and a greeting and farewell. The greetings and farewells are sometimes brief, 'Dear Sir' in one case, and sometimes are longer, as appropriate in a letter: 'To his Excellency the Lieut Governor'. Even the important letter, in which the Association informs the government in Sydney of what they have done and requests legitimization, has no detailed farewell, only Batman's signature. Yet even in this deinstitutionalized setting, the essential framing of the message remains with greeting and farewell. Similarly, the power of the signature to confirm who has written remains untouched, with a signature on each letter.

We should note the urgency and difficulty of these letters. For example, among them is a letter from a rival, Fawkner, to the government in Sydney accusing Batman of being drunk and failing to stick to existing land deals and requesting that someone urgently comes and puts the settlement in order. It is not clear who this was sent to as the letter is not folded with an address on it, and the greeting has been reduced to a minimal 'Sir' at the start. It is unfortunately difficult because of the small number of letters to judge such factors as references across letters or personal exposure. The letters fall into ones within the Association, in which the context and communication across letters is clear, and letters to outside authorities, in which case everything has to be introduced and explained. The content of messages thus also gives up some means of stabilizing communication, but given the situation that is hardly surprising for letters effectively travelling within the Association.

Overall, in a fluid and dangerous situation, which yet offered the potential for sudden wealth, and without a functioning postal system, communication was largely reduced to the three practices of framing a message with a greeting and farewell, writing a message and adding a signature, which still reflect the bare bones of transmission. Moreover, older communicative practices made a comeback in the transmission of letters by hand, though there is no evidence that this was accompanied by a return to the practice of the messengers witnessing and authorizing the communication, rather they simply took the place of the transport that a postal system normally provided.

Conclusion

The exceptional case of the Port Phillip Association and communication outside institutionalized postal systems is merely suggestive; however, it gains significance in the context of the quantitative and qualitative analysis already conducted. All this evidence taken together suggests that a number of routines were well established and that these routines provided a means of binding message to individuals, even in situations of great temporal, spatial or institutional distance. Presence was created through these binding practices.

Signatures, seals, self-identification, references to other letters, self-exposure and the ubiquity and nature of greetings and farewells are a strong set of material practices which, taken together, function to create the transmission of meaning by stabilizing who the author is and where meaning can be found. The author is performed and authorized by the creation of a body that touches the letter. This body is generated in an indicative way by the signature and seal, which provide markers that in their bald, simple statement assert that whoever is named in the signature or who carries that seal must have touched this letter. Further practices though not universal, such as referring to letters sent and received and constructing sequences of letters or of revelations about psyche or body or the telling of self-identifying stories, all reinforce the simple inscriptions of signatures and seals indicating that a body with the identity of a son, sister, mother, employer, employee or other such could be taken as having generated and sent this message.

The message itself is bounded at top and bottom by the ubiquity of a greeting and farewell which can be taken to indicate the words which carry the meaning the writer wished to transmit to his or her audience. Creating co-presence between writer and reader(s) by stabilizing the identity of the writer, then also creates the possibility of the transmission of meaning. The meaning is dependent on the prior communicative practices not in its content but in the very possibility it can be transmitted with some self-identity from writer to reader(s). For this reason, too-close analysis of the meaning of letters will obscure the surrounding practices which create the very possibility of such meaning being transmitted.

The evidence of letters to and from colonial Australia, when taken up through the eyes of communicative practices, is that the body of the writer needed to be present in the letter, even when it then travelled for months and was read by an audience thousands of miles apart from the writer. If the body is taken from these letters – if the signatures, seals, style of greetings, shape of handwriting and shape of signature, anecdotes of co-present times and more – then the possibility that this message could be transmitted would disappear. All that would be left would be a faceless set of stories that no reader could know who they came from and who could meaningfully read them. Communicative practices prior to the internet were hypothesized as authorizing some kind of co-presence through material practices and performatives that produced and connected a body to the writer's identity and the letter's materiality. Once this co-presence is created, then the transmission of messages across the globe became possible.

Notes

1 The epigraphs to the chapter are from Henry Howard Meyrick 1841 (MS 7959, 16 May 1841) and Niel Black 1841 (Mackellar, 2008, p. 213).

2 All references to Meyrick's letters in this section are to the dates of letters because all letters come from the one box, MS 7959 H15789–H15816, which would be tedious to continually repeat.

4

Virtual Worlds: Internet Communicative Practices

On the internet:
All women are actually men,
All men are actually little boys,
All little boys are actually FBI agents.

Player's signature[1]

Introduction

To explore the kinds of communicative practices, the internet has introduced means coming up to the moment. Castronova has suggested this moves so far up to date that here we have a future of immersive worlds in which we gain not just various identity markers, such as email names or the name we use to build an eBay reputation, but in which we also take on new imagined bodies with marvellous and strange abilities (Castronova, 2007).

This dream of a virtual body was for a long time, and for some still, believed to be the product of fully fledged virtual reality – in which immersion would be total and we would be able to physically run using our 'real' body such that lifting the left real leg also lifted the left virtual leg, or hoof, or whatever it had been imagined to be (Rheingold, 1991). However, instead of such a deep virtual reality, millions have already demonstrated that virtual worlds and bodies are being lived by sitting at a computer and linking to others using avatars seen on a screen that can be controlled at the individual's discretion, even if the interface is keyboard, screen and mouse rather than the imagined head and bodysuit of deep virtual reality. Such persistent virtual worlds, or a light virtual reality, lie squarely in the lineage of online communication that allows multiple individuals to connect from computers located anywhere geographically, as long as they are connected to the internet. But this standard power of the internet is given its particular form in virtual worlds by being crossed with the lineage of computer games that involve specific types of interface use and social norms and that connect to a range of cultural elements, such as the swords and sorcery myths of Tolkien and other fantasy writers, the cyberpunk legends of those like Gibson and Stephenson, the board gaming of Dungeons and Dragons, the space combat myths of Star Wars, Star Trek and many other elements of computer and pop cultures (King and Borland, 2003; Castronova, 2006). The virtual worlds created from this mix of 'light' virtual reality and pop and gaming cultures represent immersive and acute forms of internet-dependent communication.

These are persistent worlds in the sense that the world continues to exist whether the individual who has an avatar in that world is operating that avatar or not. The world goes on while the individuals who inhabit it only come alive at various times. In some worlds, avatars can be checked and when approached it is simply noted they are 'sleeping'; for most such worlds, an avatar simply disappears and any attempt to search for the avatar leads to a null result, which all other avatars and their players can interpret as someone not being logged in.

These are 'worlds' because the environments presented on the screen include many spatial markers which create the illusion of a land 'out there'. In such a land, whether it be set in contemporary times (but maybe with super heroes), in outer space or in a far mythical past, the player controls a representation of a body which can be moved around issuing commands via the keyboard and mouse. Sometimes, as in first-person view, the player sees on the screen as if they are the body in the world; that is, they may only see their virtual hands on the screen. However, in many worlds, possibly the majority, the player sees their avatar-body from a camera view perched just behind and above their virtual self. This camera view can be moved in and out, rotated and so on, with every view reinforcing and re-presenting the virtual body in the virtual world. The keyboard also connects to a range of channels through which messages can be sent privately, to groups of friends, to those nearby or those all in a particular space. Moreover, chatting through voice communications, whether integrated into the game or by running a VoIP program simultaneously, is available and increasingly used.

The worlds accordingly rely on and generate a bodily duality. There is a body of an avatar that is represented on the screen. It might be a troll, dwarf, elf, space marine, human, superhero or some other, depending on what the particular world allows, but there will be some visual representation that is customizable in variable ways and which can perform certain actions. This viewed body, or avatar, is controlled by a second body that sits using keyboard, mouse and screen to control and move their avatar. This is a body that sits, views and seemingly commands, though this sitting body is also circumscribed and produced within the technical and cultural limits of the persistent world. For example, the pains in hands often felt by gamers called on by their interface to repeatedly strike certain buttons suggests the sitting body being commanded by the interface, even while the seated body directs the on-screen avatar. The bodily nature of these worlds is thus a doubling in which both bodies are constructed and circumscribed by the duality that is entered into.

These worlds are collective, in the sense that each world contains more than one body double of avatar and controller. A key design of these worlds is to place such body doubles into contact with each other, interacting in various ways that the world enables. The interactions are communicative through text and through the avatar's bodily interaction. Further the worlds take particular care to generate a sense of space and place so that actions and communications appear to be taking place 'somewhere'. An obvious example is that in most worlds there will be a command to 'say' which will print whatever is typed as a speech from the avatar but that text/speech will only appear on the screens of others who are, in that world's terms, close by. To send a message to someone far away in the virtual world will require a different command and will usually not be shown to those who are nearby, thus implementing and reinforcing a sense of spatiality.

Finally, such worlds are 'themed' and have purposes built into them, meaning that interactions are often highly stylized in relation to the design of the virtual world. The majority of these worlds are commercially produced and sell themselves as games that allow alternate realities to be experienced, such as Star Wars, Star Trek, Lord of the Rings and so on. There are other types of persistent worlds, with the best known being Second Life. Second Life offers an alternative to themed gaming worlds in that it provides little in terms of pre-defined tasks which contribute to the story of the themed world, and instead provides a framework that allows avatars extensive abilities to build new items and to customize avatars and built environments. This lacks the progressive sense of most virtual world games which see avatars pursue preset goals and develop pre-defined abilities, whereas Second Life's purpose is to provide a self-directed second, virtual life based on building and customizing within the virtual world (Malaby, 2009). Life in Second Life is then different to the gaming worlds that the following analysis draws on, consisting as it does far more in the production of the space and the elements of an online existence, whereas gaming worlds tend to circumscribe the ability of players to affect what is persistent in the persistent world.

The following study stays with the majority of doubled bodies in some of the most popular gaming persistent worlds; for this case study, these are primarily Dark Age of Camelot (DAOC) and World of Warcraft (WoW), though ethnographies were also pursued in Warhammer Online, Planetside and Rift. The two primary worlds are both swords and sorcery themed in which characters can create spells, fight using swords or maces, wear armour and so on. Anyone familiar with Lord of the Rings will be familiar with these kinds of combinations of sorcerers, fighters, healers and so on.

All these game worlds are based, ostensibly, on combat between avatars controlled by players (player versus player or pvp) and avatars controlled by software (player versus environment or pve) or combinations of these two. In each environment, the object is to 'kill' the opponent, though 'killing' here involves the necessary and inherent right to resurrection, meaning death is temporary as a doubled body's avatar is reborn back to its pre-death state and computer-controlled avatars respawn. In such worlds, there is a clear and extensive set of tasks to engage with that offer different means of progression and achievement. This stretches from the near-universal levelling of characters, in which there are preset ways of a character moving through a ladder upward in which each step up a rung involves the character gaining new abilities and becoming more powerful. Once the character has reached the maximum level, they are then open to the final or endgame situations, which then continues other forms of progression such as gaining better equipment or opening up areas that were hitherto closed or too hard to experience.

The use of these worlds for arguments about communicative practices is that they represent a clear and extreme version of communicative practices based on internet technologies. These persistent virtual worlds depend on the internet; they could not exist without internet technologies and their wholehearted dependence should heighten the impact of communicative practices. To access these though requires, as it did with letters, a shift in focus to how meaningful transmission is made possible. As with colonial-era letters, the aim is not so much to discern the 'meaning' of the content of communicative exchanges but to try to uncover how it is such meaning can be created, delimited and passed between senders and receivers in the first place. The path to this is to take up presence and see how it is developed and which materialities and performatives are in play.

Different empirical content will be drawn out to try and suggest an overall picture. First, some smaller-scale examples from across the practices of mmpog gamers will be examined. Second, a deeper engagement with the construction of a guild in Dark Age of Camelot will be analysed. Finally, a comparison to a guild in WoW will be presented. Across all these examples, the question is one of the construction of communication through material practices and how a sense of presence is created that then allows the transmission of stable meanings between senders and receivers.

Communicative practices in persistent virtual worlds

The theory of internet communicative practices so far proposed is that communication based on internet technologies has to resolve a radical instability in identity markers and that this is done by identifying senders and receivers based on their style of communicating. The instability of identity markers has been explored in studies of the internet before, primarily implied in the extensive attention given to the idea of identity fluidity on the internet (Jordan,

1999, pp. 65–79; Baym, 2010, pp. 105–9). However, most of these analyses have been in relation to theories of subjectivity and identity rather than examining communication. It is possible that the attention identity fluidity has attracted in proposing different ideas about identity and subjectivity that are possible in internet-dependent contexts has detracted from seeing that it also has a role in affecting communicative practices (Turkle, 1996). What is needed to extend an understanding of identity in internet communication is an exploration of the interconnections between identity fluidity, the stabilization of senders, receivers and messages and the functioning of styles and whether these are the keys to sustaining communication in highly virtual environments. The elements of identity fluidity and its threat to the stabilization of communication will first be examined leading to consideration of the styles that are hypothesized as key to creating the possibility of transmission of meaning in internet-dependent communicative practices.

As has been extensively, and sometimes luridly, discussed, it is possible to shift identities while online in ways that seem remarkably fluid compared to identity outside the internet. Nearly all communication over the internet comes with an identity marker (such as my current work email of timothy.jordan@kcl.ac.uk), and it is not the lack of such markers that causes difficulties, but their ability to be spoofed, changed and multiplied. In 2011, incidents of someone being entirely different to who they claimed to be gained another example in the Syrian-lesbian blogger Amina Araf, who attracted international attention when it was thought she had been arrested and who then turned out to be a heterosexual-male American Tom McMaster living not in Damascus but in Edinburgh (discussed more fully in Chapter 5) (Addley, 2011). Incidents where whole communities and groups are taken in and believe someone to be online something that they are decidedly not offline, have been familiar from the beginning of the internet (Jordan, 1999, pp. 62–5). Such incidents dramatize the instability of identity markers in internet-based communication in spectacular fashion.

The point is that fluidity in identity markers through the ability to change quickly and easily to a range of different identity markers and to alter existing identity markers makes trusting an identity marker problematic. Even in the case of a well-established marker, such as an email address, that has been used for some time and is well known to those close to the individual using that address, there will be experiences of instability. One such experience is of what has been called frape or Facebook rape. This consists of someone hijacking someone else's Facebook account and posting absurd, funny or offensive messages as if they were from the account owner. Such a practice is abetted by Facebook's technical settings which make it easy to log off without signing out and thereby allowing someone else to start up Facebook and then find that they are logged into someone else's account. One tasteless example is a post that stated, 'Just found out I have herpes. . . . Ladies, get yourselves tested,' which was followed soon after by exclamations questioning whether this was serious, doubts that were then confirmed when someone stated they were talking to the account owner who had not posted this message which was untrue. Another was a post from a father stating that one of their three children was his favourite by far, which was immediately assumed by those responding to be fake.[2] In these and other examples, the style of the message will usually trump the marker of identity and this remains remarkably consistent across different sites of internet-based communication; whether email, the names people use in online fora, a twitter name, a Facebook page and so on. The inverse where an identity marker is not as it should be but is stabilized by a style is also true, as can be seen in phishing. Here unsolicited emails aim to represent as faithfully as possible the style of a known institution, such as a bank, aiming to entice the recipient into clicking on a link in the email because they trust the presentation of the email even where the email address is mistaken (e.g. a bank such as halifax.co.uk becoming halifax.org or halifax.net). Here style tries to overcome the identity marker of an email address that will never be quite right. The markers of identity that might seem to simply stabilize who is communicating are unstable in internet communicative practices and are regularly experienced to be so (Jordan, 1999, pp. 65–79, Baym, 2010, pp. 105–8).

Such instability gives a particular form to something all communicative practices do in the creation of a form of transmission. This is the everyday experience in which a sender, receiver, message and means of messaging are all stabilized and become repeatable. The ability to create transmission is threatened by identity fluidity, for example as seen in the practices examined in the previous chapter in which the signature and seal played key stabilizing roles, but it is an intensified problem when the very techniques and technologies of communication embed and support shifting markers of identity. This is perhaps the aspect of identity fluidity that has been underappreciated when studying the internet. While many have noted the ability to play with identities and the positive and negative aspects for psyche and personality this can bring, there has been little discussion of the endemic difficulties this poses for communication. When it has been discussed as a communication rather than an identity issue, it has often been taken as a technical problem of establishing identity to be solved by 'best practice' or a technological fix, but rarely as an endemic and general issue for communication.

The second key hypothesized component of internet-dependent communicative practices links closely to this endemic feature of the creation of everyday moments of communication and transmission in the styles of online communication. Style might seem a trivial factor, the way someone signs off their posts, the kinds of concerns they include, their use or not of emoticons and so on. Yet styles or consistent ways of interacting seem to be able to bear the weight of stabilizing internet-based communicative practices. A brief example, mentioned earlier, relates to trolling: the practice of attempting to provoke a controversy for the sake of controversy in an online forum. The official forums for discussing gameplay in WoW demand that every post be done under the name of an avatar being played in the game and every account can have up to eight different avatars allowing posting even from one game account under eight different names. For a time, it became common to recognize a troll if someone posted an inflammatory or opinionated statement with a high-level character (meaning considerable time invested) and the next comment, often posted almost immediately, either strongly agreed or disagreed with the first post but came from a low-level character. The implication most people drew was that both posts came from the same person using two different avatars on their account and generally someone will explicitly respond pointing out this is likely to be an attempt by one person to stoke some controversy by trolling (Donath, 1998). Style here is read and the instance of transmission – the reading of the signification of the posts and the judgement about who was the sender – is stabilized in opposition to the identity markers by using the style of the post. This understanding of in-game identity is not unlike Newman's argument that assuming a gameplayer should be identified with an in-game character is an oversimplification and that instead players accounts of their playing 'are structured around action, around environment, around activity' (Newman, 2002). In encountering online communication and styles, we encounter not so much identities as activities (Tronstad, 2008).

Here is one small example of communication and style from my ethnographies of WoW. The following is an oft-repeated set of greetings between two long-term members, and real-life friends, in the guild AS. The character N[3] is also the guild leader and the exchange occurs if the character Llirk (who is one of my characters) is already online and N then appears. The prefix [Guild] is usually in green and all text after it is also in green, this tells players that these messages appear only to guild members, with other communication appearing in different colours.

N has come online
[Guild] Llirk: Lord N <bows>
[Guild] N: <grumble>

Llirk here self-represents by typing and sending to the guild 'Lord N <bows>,' which is understood as Llirk saying 'Lord N' while performing the action of bowing. This mocks N as his Lord, something that will be undermined at every linguistic opportunity in guild life. N then self-represents as grumbling at the greeting, a reaction conveying he knows that he is being mocked as the guild leader. A different but also common response from N has been:

[Guild] N: <raises eyebrow> Someone wants something

The use of <grumbles>, <bows> or <raises eyebrow> draws on role-playing conventions in which an action can be conveyed in text by being put into angle brackets, a stylistic resource first developed online in text-only persistent worlds such as Muds (King and Borland, 2003). This is necessary because N, Llirk and all the guild members will not be in the same location in the virtual world and so cannot actually see Llirk bowing but all will see the green guild chat text. Styles thus need be read in relation to wider cultures than just those immediately generated by participants in acts of transmission. For example, not only does this example draw on wider conventions of role-playing but because these two repeat this interaction, often in exactly the same text, they also create a style or convention that goes with the interaction of these specific characters. Llirk in this way builds up a mocking style and N is able to respond also building up a style that then becomes part of repertoires of interaction they will be able to draw on. Others who know them will recognize this interaction and variation from it because they are gaining the resources to read the styles of Llirk and N.

What is being hypothesized is that in the face of a radical suspicion of identity markers the ability, as seen in pre-internet communicative practices, to attach a body to an identity has to give way to forms of stabilization that do not rely on bodies. Instead, the doubled bodies of gameplayers seek out styles of interaction that both draw on wider cultures and identify specific individuals. These interactions will be everyday moments repeated such that they create particular practices, just as signatures, seals and greetings and farewells were both everyday moments and repeated such that they became generally understood in pre-internet communicative practices. These practices will create presence and in doing make the transmission of meaning possible.

To now take these ideas further and immerse them both in virtual worlds and in ethnographic material, a detailed case study will be presented of the nature of two guilds in different persistent world games: Nyd in Dark Age of Camelot, 2001–4, and AS in WoW, 2004–11. The focus will be on the ways the guilds were located at the intersection of individual styles, local styles and wider cultural styles, with each style here understood as the repeated use from a repertoire of particular forms of interaction.

The ethnographies of both guilds involved extensive participant observation combined with review of films of in-game events, logs of chat, screenshots and several round-table discussions with guild members of the resulting ethnographies. These guilds and the game are part of a community that can be participated in as a member in the fullest sense. The methods employed to conduct this ethnography are not standard participant-observation theories but draw mainly from Kaminski's inversion of participant observation to argue for a method of observant participation (Walsh, 1999; Kaminski, 2004). Kaminski's ideas cover situations in which ethnographies can be developed by a researcher from the researcher's full immersion and participation in the group being studied, which is not always possible or desirable when conducting participant observation. For example, Kaminski's own ethnography was conducted in Polish prisoners in which he had become a prisoner, he was therefore not so much observing as able to be a full participant in being a prisoner. When the researcher is able to participate fully in both the culture from which they base their research and the culture which they are researching, then a particular kind of ethnography is possible. This was the case as I was an mmpog gameplayer before I began ethnographic research into mmpogs. There was accordingly no 'outside' position from which I could base research, while at the same time I had the advantage of full access and deep understanding of what was occurring. As an observant participator, the following may lack in some objectivity but cannot be faulted for access or richness of observation (Abu-Lughod, 1988; Kaminski, 2004).

Nyd-Mid-Pryd

Dark Age of Camelot (DAOC) was a medieval swords- and sorcery-themed virtual world whose primary attraction was the ability to participate in player versus player (pvp) realm war that pitted large numbers of players against each other. There was also considerable player versus environment (pve), some of which was necessary to compete in pvp (e.g. to gain equipment) but much of which could be played for its own sake. However, it was the pvp which was the primary theme of DAOC, which launched in 2001 and though it remains operating at the time of writing (mid-2011), its numbers have been extremely reduced since the launch of WoW in 2004. It reached a mainly US and European subscription base of 250,000 at its peak, which fell of sharply since 2004 to a 2008 number of around 50,000 (see www.mmogchart.com). DAOC featured three separate realms of Midgard, themed on Nordic myths, Albion, themed on a 'post-Arthurian' Britain, and Hibernia, themed on Celtic myths. Each realm had its own versions of the usual archetypes of players: magicians who deal high damage but have low health, tanks who deal low damage, who attract monsters and have high health, assassins who can go invisible, deal high damage and have low health, healers and so on. Each realm also had its own races, for example, Midgard had trolls, norse, kobold and dwarf, with later additions of frostalfs, valkyn and delfrang (minotaur). Players can play solo, in both pve and pvp, but more often co-operated in groups of eight which could be coordinated through chat with other groups of eight. At times, hundreds could be in the same area and this kind of large-scale pvp between three different realms provided much of the unique attraction of DAOC.

DAOC was launched first in the United States and later with specific servers for European players. I joined the European launch of DAOC after having tested it by playing it during open beta (the testing phase when anyone was allowed to play for free to try and work out bugs or problems in the game). It was my first virtual-world experience. While in open beta, I noticed in a related online forum someone posting asking whether anyone else was interested in a guild for 'mature' people. I contacted this person A, and we met in game, agreeing to set up a guild based on role-playing that we called Nyd. The guild quickly attracted one other person E, and later in its life gained three other regular participants (C, T and N, the latter being the same N who Llirk likes to call Lord N in WoW) and some other irregulars (often people who had met guild members, had active avatars in another guild, and placed one of their characters in our guild). Nyd was accordingly very small, compared to large guilds which often had over a hundred active members. It participated in alliances with other guilds to experience some of the events that needed larger numbers.

A key initial point is that Nyd was a role-playing guild, not a common or usual choice for DAOC guilds. Following the terminology of the doubling of bodies, role-playing can be understood as an attempt to deny the existence of the body sitting in a chair at a computer screen while asserting the vitality and singularity of the avatar appearing in the virtual world. However, even within role-playing, Nyd developed a particular approach. I will distil from this rich and varied interaction in a virtual world, going back over several years, a number of tenets that outline what I think of as the Nydian way. These are not rules we defined and acted by but, looking back on the ethnographic records, there are number of key components of how we existed in DAOC that underpinned and created communication. There are three general ways of being or tenets of Nyd that I will discuss: no smileys, stay in game because Midgard is real and enjoy, it is not Tolkien. I will touch on these in turn, but it should be emphasized at the outset that they are abstractions of something that simply grew between those of us in Nyd. It should also be recognized as only one way to do the thing called 'role-playing'.

No smileys!

And no abbreviations; none of those things so beloved of those of us dealing with too much text and typing in our lives. This is the most straightforward of the three tenets. One of the first times A and I headed together to the area called Emain, at that point and seemingly forever the main realm-war area, we had the following exchange:

[Party] A: 'Thing happen quickly here, so it's ok to use abbreviations'
[Party] Krill: 'Ok, m8'
[Party] A: 'Aaaargggh not that one'

Abbreviations such as m8 are familiar to many from texting on mobile phones. In this context, it was simply the abbreviation most likely to annoy A. In addition to text-message shorthand, players in DAOC drew on a whole range of already-existing emoticons and abbreviations developed from 30 years of text-based internet communication: lol for laughing out loud; rofl for rolling on the floor laughing; :-) for a smiley happy face; ;-(for a winking unhappy face and :O for a surprised open-mouthed face. All these forms of shorthand allow people to indicate how their text should be interpreted. The message 'idiot :-)' from a friend should probably be taken differently to 'idiot :-(' from someone else depending, of course, on what idiotic thing you may have done.

In addition to these linguistic resources, the game itself produces a language which derives both generally from multiplayer online gaming (afk for away from keyboard, for example) and from things specific to DAOC: amgO means Albion mile gate Odins, a place; ml1.5 means master level quest one, part five; ooe, oos, oop for out of endurance, out of speed, out of power. Other games have similar abbreviations. In WoW, I learned of AV meaning Alterac Valley, mc meaning mind control and so on. There are many of these spread all through the game that are gradually learnt by players and some abbreviations cross games.

Nyd set its face against all these. We spoke in full sentences, we hardly ever abbreviated, though in the heat of combat we sometimes did. Our exceptions came from necessity. If I was killed at the Albion mile gate Odins by a full group and I wanted to quickly call in some nearby people to get revenge, I could type: 'albion full group, at albion mile gate Odins, rebuffing' or I could type 'amgo alb fg rb'. The time might seem minimally different between these two but in such a case revenge demands the quickest message possible. However, overwhelmingly we did not do this. Largely we gave things their full name and while it hurt my typing fingers, it gave us a distinct presence.

Distinct enough that when we had made friends with others outside our guild who enjoyed our kinds of interactions, we would sometimes be asked advice on how to discuss things, particularly if there were eight of us in a group together. On more than one occasion, I received a private message that no one else could see from someone outlining what they wanted to say and asking how to do so within our styles. For example, someone had a difficult internet connection that was giving them lag and hence slowing their in-game avatar making them react in a delayed way, and they wanted to let everyone know without breaking the atmosphere created by not using abbreviations. As this was something we had had to do within our guild several times I already knew that we talked about the 'Gods of Connection' and how they were persecuting someone.

Sometimes we articulated this style among ourselves when we jokingly used shorthand. I remember with A spending sometime ambushing Hibernians in Cruachan Gorge. This, until a total revamp of the realm-war zones (called New Frontiers), was the main entry point for Hibernians to defend their frontier lands. We had snuck all the way there and hid behind trees, sprinting out behind any poor sap we felt we could beat and belting them down. It was at times tense, at times too quiet, though huge fun when it worked. We were both dressed head to foot in green, something that had become a Nydian trademark. I played also a very large character which could be a bit of a shock for someone suddenly turning and finding me right behind wielding a two-handed hammer. On one occasion, we completed our murderous mission perfectly. A firbolg appeared running down the path, we sprinted from the trees directly behind him, he did not notice till A mesmerized and then stunned him, while I hit him. He was already low health when he finally turned to face the not-jolly green troll right behind him.

After we had hidden again, I told A:

> [Party] Krill says: 'I saw surprise in his eyes as he turned around. This damned Hibernian looked at me with shock and disbelief'
> [Party] Krill says: 'and, as he crumpled to the ground, I heard his last words'
> [Party] Krill says: 'he said: wtf!!!??11??'

The dissonance of using the common abbreviation 'wtf' along with the common overuse of exclamations (themselves poorly done as the 11's show up when someone fails to keep their caps key on and so does 1's instead of !'s because they share the same key) while also using more flowery, role-playing language exposed our choices of language in a search for humour. This breaching underlines the boundary we were creating by normally using if not the more staid role-playing language at least refusing to use the abbreviations the world around us abounded in. This example also introduces the next tenets in our commitment to the real world and our more particular role-playing style aimed at humour rather than authenticity.

Even with one tenet outlined, we can see how a style is being developed that references the dominant style around us by our difference. These styles draw on resources, a key one being the use of abbreviations which we refuse. Having dropped the particular resource of abbreviations, Nyd was able to then use language in ways that would identify from particular common styles both messages and individuals.

Stay in game! Midgard is real

When A and I met in real life (irl), we had been communicating together for several hours on most nights for over a year, and until a couple of months before we met we had hardly ever talked about who we were outside Midgard, about what we did or anything else from the non-DAOC world. We knew both a lot and a little about each other. The same went for most who passed through

Nyd; for a long time, we simply did not talk much about the non-virtual world, instead treating the in-game world as real.

When we were grouped with others who did not feel the need to believe in Midgard – all the vast number of players for whom the idea of suspending disbelief would have only been met by a succinct 'wtf?' – we tried to maintain our attitude. At times this set us apart, at times it caused laughter, sometimes others joined in and sometimes others found us annoying. The simplest example is time in the game. In Midgard, the day passed in an hour, there is night, day, sunrise and sunset all in the real-world space of 60 minutes. It follows that real-world time is usually different to in-Midgard time. Someone arriving at 9 a.m. real-world time can be at 10 p.m. game time (for a long time you could check the in-Midgard time within the game but not real-world time). So arriving to a group or guild can elicit odd greetings as people say 'Morning' or 'Evening' and can be referring to different time zones – virtual and real. Nydians in a group with someone who arrives and says 'Morning' when it is dark on the screen would most likely question the sanity of the person, pointing out the absence of sun and light, often to the bemusement or hostility of the person being corrected. This was more pointed in DAOC than in WoW as the latter's in-game time follows real time and, accordingly, as long as a player is in the same time zone as the server they are playing on, it will be night in game when it is night out of game (in addition, the change in light in game is not substantial).

Nyd's commitment to staying in Midgard, and escaping any irruption of real life, was supported by a semi-elaborate language in which we referred to non-virtual lives as if they were part of Midgard. Allowing the real world in but translating it into Midgardian terms meant we could to admit the inescapability of the doubled body but also hold its effects at bay. From the start, Krill referred to having a partner, and once he had taken the surname of Iagdtod, he usually referred to her as Madame Iagdtod. I(he) also referred to the small trolls back in the cave, it being assumed trolls tended to live in caves, when I needed to refer to my real-life children. Particularly in the first year, I was interrupted a few times by a child waking and would simply say I needed to go back to the cave. The others developed similar strategies. A few times A came in excited by prospects of a new job or promotion offline and then he would talk about prospects in his day job at Jordheim (the in-game capital city of Midgard). The extent of this ability to translate the real world into terms that made sense within our virtual realm was limited only by our imaginations and typing skills.

As such strategies suggest, Nyd's attempt to escape the doubled body for the single virtual body is impossible. We were not, literally, running around the land of Midgard as kobold, norse and troll, healing and firing lightning bolts. We were, literally, sitting at computers connected via the internet while directing a representation on the screen of a kobold, norse or a troll that was healing or firing lightning bolts. This meant that however consistent we were in our language, the doubled body would reassert itself. When in groups with people who were not Nydians, it was often necessary to say things both in our style and in a clearer (to outsiders) way. This led to the rise of the square brackets. Within square brackets, we would say things using abbreviations or speaking clearly from the real world. This is a typical one:

> [Party] T says: 'I must rest a few moments. [afk 5 minutes, phone]'

And for me, there was this memorable one:

> [Party] Krill says: 'AAArrffggh must go [wash machine floooded]'

And it had, there was a lot of water sloshing around in a room instead of sloshing around inside a washing machine. Such linguistic devices managed our time in Midgard in ways that maximized the sense that we were simply there. We tried to naturalize the most unnatural world of DAOC and that made us as virtual as it is almost possible to be. Yet these examples both stress the lengths we went to in maintaining the illusion of Midgard's reality, the square brackets were brackets of last resort, while confirming that the doubled body could not be escaped.

One further aspect needs to be teased out about claiming that Nyd treated Midgard as if it were real. Staying in game is not that hard and many gamers do it, but the way we did it related to the meaning of the world we were in and not to the mechanics or workings of the game, which themselves can be absorbing. Many gamers were online for longer periods than any of Nyd ever were. I have dropped into Midgard in a real-life morning, left and come back at the end of the day to find someone has been there all that time and was intending to stay most of the evening. Such players were dealing with similar issues as we were in Nyd concerning how to manage virtual and real selves, but were taking an entirely different attitude. They often focused on the skill in their selves as players by concentrating on understanding and utilizing game mechanics. The point here is that among gamers, Nyd was not unusual in becoming enmeshed in the game world but we were unusual in how we understood ourselves in that world. For us, staying in game was a style that separated game mechanics out and focused on the meaning of the world we were in. Many others were just as immersed in the virtual world of DAOC as we were, but their focus was usually on the way they could play the game better or worse in terms of the game mechanics and not in terms of the meaning of the world. And, to finally complicate the picture, it is also not that we in Nyd were unconcerned about how well we played, but we refracted that concern with game mechanics through our role-playing.

This also placed a premium on our way of playing, not only were we sometimes hard to understand for other players, but also we had to type considerably more and make the mental effort to think of ways to describe things, rather than simply going with dominant and oft-shorter ways of interacting. In some contexts, when things happened quickly for example, our way of typing was simply worse and led to us being worse than we could have been: for example, saying 'I'm being hit' can be enough of a delay over 'aggro'[4] to make the difference between being killed in an encounter or killing. Encounters sometimes need swift changes and the emergence of voice communications certainly made our typing a greater handicap. This was particularly the case in pvp which engages the uncertainty of what a human might do as opposed to an avatar controlled by software that will have pre-scripted moves.

This premium led many to be puzzled with why we bothered or annoyed if we tried to use our normal practices when with them. Sometimes grouping with random people would lead to their frustration and anger. On a few memorable occasions, this led to them leaving and later getting in touch having gone shouting to others how stupid we were but then being told how we were well known and deserved their respect. Our particular identity built from these communicative styles was in these few cases revealed as something understood Midgard wide by longer-term players and even sometimes liked by people who had been around us long enough.

Enjoy! It is not Tolkien

While the previous tenets are close to many role-playing ideas and might be found quite widely in role-playing groups, Nyd also developed a niche within such styles. A classic role-playing style was to think out a personality with a backstory and then play your alter ego accordingly. For example, someone might behave as if they hated dwarves irrationally because, in their backstory, dwarves had failed to save their brother. This led to a kind of interactive fiction which role-players can develop (MacCullum-Stewart and Parsler, 2008).

We in Nyd did not pursue this style. We sometimes had stories which filled out who our characters were but we often did not bother very much with them, and they often functioned more to explain bits of the world and our interactions than to construct convincing personalities. For example, as already mentioned, I had small children in real life and if my attention had to be taken up by them when I was playing, I would mention that the 'small trolls' back in the home-cave. In this way, I was generating a backstory of a troll family life but doing so mainly in order to be able to stay in Midgard but communicate important issues about the non-Midgard world. We talked sometimes about setting up role-playing 'events,' but we hardly ever did this (though others did and we participated). Rather, we tended to use being in Midgard as a basis for fun through verbal twists and turns, using the environment to create situations which made us laugh or engaged us. We hardly ever role-played in the following style:

> [Party] Elbereth says: 'Battle awaits! The ashmonger mocks us from a distance, let us call him to us and cleave his skull open with our blades.'

This is not an uncommon way of role-playing with roots deep in the kind of language of fantasy or swords and sorcery fiction. The response from within Nyd to Elbereth's statement did not however stay within such styles.

[Party] A says:	'He calls you names? What kind of names?'
[Party] A says:	'Long-winded perhaps?'
[Party] Krill says:	'Me got only hammers, no blades. <looks sad> I watch this one then.'

In this case, it was then followed by a discussion of what the appropriate names for Elbereth might be. In the meantime, we would almost certainly cleave a few skulls of ashmongers, all the while avoiding the impulse to say 'lo' or 'behold' too much. There is, of course, nothing wrong with role-playing as an Elbereth might, it simply was not our main way. For example, we admired the dwarf we came across one day who had made shouts in the dwarvish language Tolkien had invented, but we were not tempted to speak in this high-minded way ourselves.

Another example of this is the use of troll-speak. This, like a form of kobold some players in the United States developed, is a speech specific to a certain race. This derived from the general view that trolls were not only somewhat stupid, based on them being the largest and strongest characters in the whole of DAOC, but were also the slowest and least dexterous. Here is one short example of this talk. It is from Dl and Da's in-game wedding, which I presided over as their priest. Dl yelled to be sure those who are gathered around (some 20–30 others) could all hear.

> Dl yells, 'Me want to fight by yur side and lean yu my shield in da battle and help yu bashing da enemy and make our surname to be feared in da realm of da enemy so we get to known as da big troll Dl and his small but bootifull kobolt wife.'

But Krill, though a troll, never spoke this way, but stuck to the full sentences of Nyd. He would also fairly constantly point out that trolls were not stupid, just slow and strong. I had a role-playing backstory ready for this, as early on I had decided that despite being a troll I was not able to stay in-character talking trollish. This story was about having been befriended by a norse (one of the other optional races of Midgard) family who had taught me my letters. This family had lived unwisely near bloodfelags (monsters in Midgard whose story is one of racial purity, several quests set out that the bloodfelags' believe the norse should rule over other races). Krill was seen by bloodfelags visiting the norse family, who were consequently one day attacked by the bloodfelag. Krill happened to arrive in time to fight off the bloodfelag, only to find the parents dead leaving behind two norse daughters. These daughters Krill adopted and brought up as his own.

The story allowed Krill the luxury of not using troll-speak as he had been taught Norse or common language. I thought it would serve to manage the doubled body because both my daughters had seen the game I was playing and were interested in it. I had allowed them to create characters they used to run around the world, and both wanted female characters. My proposed backstory allowed me to say these norse women characters were my daughters and blur the online and offline, also allowing me to explain why these characters were not allowed to join groups or guilds. However the key element, for understanding Nyd, of this elaborate story and its management of doubling is that I told this story precisely once in years of playing. Though I had worked the story out in my head, had even been to bloodfelag areas and identified a house they did not occupy but was near to them which could be used as the site of the story, and I had expected to use it to explain how I was playing, it simply did not come up. I was rarely asked why I did not speak in trollish from either within or outside the guild, nor were there many occasions when we told each other stories to develop our interactive fiction. All this marks how such backstories were not core to our communication and identity. This illustrates the way that backstories in Nyd tended to be used to allow us to talk in the way we did, creating jokes and playing with language, instead of the more familiar role-playing methods for creating a character's 'motivation'.

Nydian communicative practices and AS communicative practices

The tenets I have outlined illustrate some of the main lines we pursued, some of our habitual ways of acting so that we could communicate. These were not the only ways, for example, at one point we started dyeing our armour green. This may have come from time spent hiding in trees and green landscapes, but it then gained ground when someone insulted one of us for looking like snot. Having developed the kind of identity we had, this immediately led to all of us going green and claiming it as a guild trait, to the point of accusing others of stealing our colour and telling them not to be green. Yet even such a visual way of identifying ourselves was at least partly in the service of creating interplay based on verbal dexterity, humour and some provocation.

The style of Nyd has been presented far more neatly than it was lived. We did not design these rules and then live by them, neither did we write them down as rules that anyone wanting to join Nyd had to read and sign up to. No doubt other Nydians would describe the style with different emphases. Rather what I have called the Nydian style grew as part of our gaming. It is only looking back at the ethnographic record that these three tenets seem to me to encapsulate how our communication worked.

In terms of communicative practices, Nydians developed a style which identified us to each other and to those outside our group. The focus on this particular style is asymptotic of most mmpog gamers, because we symbolically rejected the doubled body by banishing the offline body. We then created our forms of transmission by constructing particular ways of interacting out of this choice. The styles have been abstracted and summarized in three tenets, and it is following these tenets that formed a style that allowed transmission of messages between us. As noted early in this chapter, styles are always going to be somewhat collective in the sense of resources being available to a number of people whose use of these resources then creates a personal style which identifies an author/recipient and their messages. We can see in the three tenets a summary of various resources those in Nyd could use to mark out a style, within which they would then begin to mark out their personal style; for example, in Krill's choice not to speak trollish and the early guild choice to speak in full sentences eschewing abbreviations. Nyd pushed these styles to the fore as we were able to create successful transmission of messages, many many messages, between avatars whose identities was fundamentally constructed within these Nydian tenets even though we had disconnected the seated body from these identities. The example of Nyd as a guild suggests both that it is possible to create presence and thus the possibility of transmission, by using characteristic styles of communication that authorize identities and associated identity markers. A Krill who logged on and started speaking liberally using 'lol, wtf, rofl' and so on would have been under deep suspicion within Nyd for simply not being me.

However, Nyd was a somewhat unusual guild, very small numbers combined with a distinctive and strongly held ethos, suggesting that perhaps these conclusions might not hold more widely. If Nyd allows a focus on the creation of presence, the body and styles in internet communicative practices, because Nyd was doubly extreme in its virtuality, because it exists in a virtual world and extreme in its fight against the doubled body that sustains virtual worlds, then will the results from examining it hold widely? To touch on this, a second guild, this time from WoW, and its differences to Nyd's three tenets will be examined.

Several members of Nyd would eventually merge into a new guild called AS when WoW launched in 2004. By this time, Nyd had already closed in DAOC and Krill had joined another guild there, though most of his friends had already tired of DAOC and moved on. Some friends, such as E, had left gaming entirely, not sure that the time they had spent there was worthwhile, some had moved to other games in the interim before WoW and others had had a break from gaming in anticipation of WoW. The launch of WoW brought a new guild with many we had known and been friends with in DAOC, and most Nyd members who were still actively gaming joined up. By this time, a number of people who had played in Nyd or been close to it had become real-life friends and all of these found their way to AS. This guild began with a light role-playing ethos, certainly not as strictly held as Nyd's, but gradually moved away from role-playing entirely. The difficulty of maintaining role-playing in the face of a general culture indifferent to it is aptly caught by analyst of virtual worlds and gaming Tim Burke (Hagström, 2008, pp. 280–2).

> I do think that making 'interactive fiction' can be a marvelous kind of jazz, but that it's hard to do when you're Egil Skallagrimsson and you find yourself fighting alongside Battlechimp, Long Dong, Dr. Evil and Osama bin Laden. That's a discordant note in interactive fiction, much as reading Lord of the Rings and seeing Gandalf and Aragorn go off to Mordor with Mr. Poopypants, Spiderman and ShadowDeathLord might be.[5]

We can gain some perspective by considering the three tenets of Nyd in relation to the later guild AS. The first was no emoticons. This did not hold in AS where in 2011 all kinds of emoticons were in use; including the unfathomable :P, which is sometimes held to be a tongue in cheek, sometimes to be poking a tongue out, but seems (rather like lol) to be used at the end of nearly any sentence and to therefore be close to impossible to understand. The following perhaps demonstrates the change.

> [Guild] [W]: I tried shouting they were cheap and sold them >> <<
> [Guild] [Llirk]: What does >> << mean?
> [Guild] [W]: Shifty eyes ;P

I had seen the shifty-eyes emoticon a few times but never really understood it. Yet even with the use of such emoticons, and lol and rofl and so on, AS retained a greater amount of full sentences and proper punctuation than can be seen in most general chat or when grouping with others. This is really an emphasis within what is a more general WoW culture and marks out AS without disconnecting it from such culture.

The second tenet of treating the game world as real and the offline world as unreal or a shadow was clearly not part of AS's communicative practices. Yet again, AS practice was marked by its past and the kinds of players it has attracted and managed to keep involved one way or another across a series of games. This then turns on various communicative practices which invoke the past but are refreshed for a different type of guild identity.

There is no question that AS in 2011 was radically different to Nyd or to several of the other guilds it grew from and much closer to more general gaming conventions. It was common to find references to people's outside lives, to discuss gaming mechanics, to talk about the gaming environment and changes to it and to the company that owned and ran WoW and more. Alongside these, interwoven, might well be comments made as if the in-game world were real and using role-playing conventions; for example, the interplay between Llirk and N outlined earlier. But such irruptions of role-playing contrast with Nyd in that they do not form core dynamics but instead exist simply as another resource through which particular styles might be generated.

The third Nydian tenet was that role-playing was far more for jokes and interplay rather than for character building or generating interactive fiction. Here, cross-pollinated with conventions drawn from players from other guilds who joined AS, is something of the communicative quality that Nyd developed. It is changed in that there is little chance of the more formal or traditional role-playing that develops characters in an interactive fiction and hence the choice to focus on what wordplay can be generated is now contained in relation to wider gaming cultures. Here the guild banter, not unlike any group situation such as a sports team, is marked by the occasional use of role-playing devices instead of being assumed to be within a role-playing environment.

Comparing the three tenets of Nydian communication to AS we find some continuity but mainly differences. Within these collective sets of communicative resources each member develops a particular style – as seen with W's use of the shifty-eyes emoticon and Llirk's confession of not understanding them – which allows people to identify them. This should be seen in the context of an AS practice that was nearly unknown in Nyd, in which people have several avatars that they play and may swap quite quickly between them. Even my account which, while having a clear main in Llirk the warrior tank, had a second avatar at maximum level that I brought up because of their skills in collecting various resources available in the game. In quiet times, where there is nothing that can be done meaningfully on Llirk, there was even a third character, a mage who fires magic bombs, who also finally grew up. And this is quite a limited roster, with many who have considerably more alternative characters. It is then not uncommon for individuals to complete one task on one avatar and then log off to log on a different avatar and take up another activity. In ten (or more)-person raids when a particular fight is not going well, it is not uncommon for there to be a discussion of whether a few people can change characters to give the raid some different abilities that might help.

Amid this shifting set of identity markers, as various names appear in the guild roster that are all familiar but which can be hard to attach to a particular body or a particular main avatar, it is the style of individuals that remind members of who they are communicating with. W's use of emoticons and characteristic concerns will prove powerful in reminding everyone that another avatar with a different name is being controlled by the same offline body allowing communication to be stabilized. Communication here derives from the use by individual players of collective resources available from the game world and from the nature of their particular guild, such that once particular styles are recognizable then various avatars can be stabilized as being particular senders and receivers of communication even as the identity marker radically shifts.

An example of this was the case of a long-time guild member D, who began suddenly behaving uncharacteristically, instead their new actions matched the style of someone who had their account stolen. D could be seen taking out resources from the guild bank, a kind of virtual storeroom that guild members have access to, and selling them on the auction house. D's activity was noticeable as they were taking out the maximum number of items, itself unusual, as soon as they could, and then a few noticed the same items on the auction house. Once this activity was repeated a few times, D was kicked from the guild and a notice about why was posted in the guild forum, an out-of-game bulletin board for guild members. The forum thread was entitled 'For D, reason for kick' and the first post was from a guild officer stating:

> I have kicked D from the guild as I think he has been hacked, as I can see he has been emptying the Gbank. Just contact an officer to get reinvited when you have control of your account again mate :)

Subsequent discussion in the thread concerned how to contact the offline D as it was thought he did not frequent the forums. It was also noted that his character had been stripped of all its gear, confirming him having been hacked and his account stolen (hackers usually sell all they can to generate in-game gold that they can transfer to other accounts). D was offered a re-invite when two things were noticed. First, that he had all his gear back on, which was standard practice as the owners of the game usually returned stolen or sold equipment once a hack was confirmed. And second, a long-standing member of the guild posted in the thread.

> I've been chatting to him. Its D :)

The identity marker was again here trumped by the style of interaction, in this case quickly trumped because the new style was highly unusual for D and was recognizable as a character having been hacked. The guild re-invite was issued after someone confirmed the avatar was back in D's offline control simply by chatting to him in text, confirming D's identity by his ability to communicate again in D's style.

We see a quite different style of communication in AS to that of Nyd, for example AS's easier connection to more general WoW and gaming cultures, rather than to role-playing cultures. In both cases, the presence that is the basis for communication is generated by stabilizing identities and messages based on characteristic styles of interaction which draw on and develop repertoires for interacting. This can finally be underlined because in both guilds it is not that stable-identity markers did not emerge, but that where an identity marker became solid – such as my name of Krill – these remain authorized on the basis of the style of communication that this identity-marked avatar emits and receives. The case of D being hacked confirms and underlines this. Identity markers remained unstable, an example of which is my failure to obtain the name Krill when starting in WoW simply because some other player was able on launch day to log in before me and create an avatar named Krill. My reaction was to reverse my name and play for many subsequent years as Llirk, even though long-term friends sometimes still refer to me as Krill in game which occasionally confuses newer members. This shift of name was repeated when WoW was abandoned for Rift, I initially gained the name Krill, but when the guild transferred servers to a new shard I found Krill was taken and I had to revert again to Llirk. We were not only aware any of us could at an instant put someone else in charge of our avatar but also we were in a context where people might let others play their avatars or had multiple avatars and so on. This ethnography suggests that viewed qualitatively, it was style that created identity and not identity markers that enabled styles of communication. Not only do Nyd and AS demonstrate the ways styles work and the rich resources they draw on, but they also confirm that the instability of identity markers is a well-understood context for internet communicative practices and that this means that such markers are subject to confirmation by styles of interacting.

Styles and styles

The style these guilds developed, which themselves drew on wider resources such as role-playing conventions and online-gaming conventions, have been the primary focus as they test the case for styles creating stable message transmission as clearly as possible. It was not only possible to develop a durable and substantial set of interactions between identities whose stability was based on repeated types of styles, but it was also entirely successful. Nyd gained a significant reputation even based on such a small group of players within Midgard-Prydwen (which we should not overestimate in number but was likely at any one time to be somewhere around 750–1,000 active players). This can be illustrated by recalling 'Nyd's Last Stand'.

This was an event when Nyd had dwindled to being an even-smaller guild with A and E having left the game, though still in contact outside Midgard. This left two to three players coming online and much of the style of Nyd feeling to me like a weight which held me back from being able to game in a way more consonant with the general gaming style. I therefore decided to leave Nyd and join another guild full of people I had met and liked but which was not a role-playing guild and which had considerably more members, making it, I hoped, easier to find others to game with. After some discussion in which I argued that Nyd should close down, it was agreed to hold a final event and shut Nyd's doors. To that end, we organized an attack on Albion, with the aim of taking down one of their castles and defending it. We also allowed anyone who wanted to join Nyd for the evening to do so on the conditions that they left at the end of the night and they role-played as we normally did in Nyd. In a turnout certainly larger than I expected, over 70 Midgardians joined Nyd, most of them then dyeing themselves green to fit in and doing their best to stay in character and another 30 remained in their original guilds but joined in the fun. We ended up taking a castle, having it attacked by over a hundred Albions, killing them and venturing deep into their territory. What it showed was that many had seen our style, whether superficially in wearing the green or deeply in one player's injection of a long-term role-playing story he was developing into the evening (a story that involved as part of the raid, a feud and duel between Krill and another player R and would later be more fully explained by a work of fiction the player wrote that included these events). That evening demonstrated not only that we had a style but also that our style was recognized widely across Midgard.

The Nydian style here, built from the ways we interacted, gave us an identity and confirms, even within an unusually small and distinct group, the ways styles can form a basis for communication even when the body and identity cannot be used to confirm who is communicating.

Like the previous chapter, this discussion has had to shift focus from the more usual analysis of the content of messages in order to try and bring into focus what it is that creates the possibility of transmission. Letters are more commonly analysed for their content and what that tells us about individuals or the times they live in, similarly online games are often, and importantly, analysed for their playability and the identities and socialites they produce. However, rather than analysing the form of communication and how that allows a meaningful written message in the first place, this ethnography of online gameplaying has moved to trying to see what is it that allows the stabilization of authors, receivers and messages in environments where the physical body may be either absent or an unstable and unreliable indicator of identity.

I have argued that these ethnographies identify styles, or repeated types of interaction, that generate a stability to the roles needed for the transmission of meaning: the sender, receiver, the message that retains its identity while it moves and so on. Interactions were here found to grow from various technical (text, avatar movement) and cultural (role-playing, online gaming) resources that were inextricably intertwined in the fluid and dynamic environment of a gameplaying persistent virtual world. Many of these interactions were drawn from what we might call collective repertoires which provided players with ways of building their individual identity while at the same time locating themselves in relation to other individuals and collectives. Interestingly, it has in these ethnographies not been possible to clearly separate out the group and individual interactions instead noting how located in particular contexts individual styles emerge as part of drawing on and contributing to collective resources.

It might seem that an online persistent world offers up many identity markers. The names of avatars is one seemingly obvious and powerful such marker. But I have argued that such markers remain authorized by styles of interaction rather than the other way round. This can be emphasized by noting that even an avatar's name, appearance, race or gender are no longer permanent in WoW because this game now offers paid services which allow a player to change all these markers on their avatar. What remains is the way a player plays and from this flows the possibility of enunciating a message and that message being understood.

Notes

1 Seen on official WoW, Euro forums, 11 October 2010.

2 These were drawn from www.bestfrapes.com, accessed February 2012, but any search will turn up multiple sites offering similar examples.

3 Most player names, except my own, have been anonymized by being reduced to a single letter. My avatars were named Krill or Llirk. Some random player names of people who were only met once have been altered but kept as names. I have not anonymized Nyd's name as the guild was closed down over seven years ago; however, I have anonymized the name of the guild in WoW that I was part of.

4 'Aggro' would have been understood as that monster is aggressive to me and is hitting me, usually used to make sure that the person who has aggro is the right person.

5 I take this quote from an online discussion with Tim Burke, thanks to him for letting me use it.

5

Internet, Society and Culture: Anxiety and Style

Anonymity is authenticity. It allows you to share in an unvarnished, unfiltered, raw and real way. We believe in content over creator.

Steve Poole/moot Founder of 4Chan[1]

Two stories: Julie and Amina

Two stories illustrate communicative practices after the internet.

Amina Abdallah Araf al Omari (often referred to as Amina Araf) presented herself on a new blog in 2011 as a lesbian Syrian-American telling her story of living in Syria. The blog attracted attention for its insight into life in Syria as a lesbian, this was particularly due to high interest in the Mideast because of the revolutionary social transformations then occurring there. A flavour of her writing can be gained from part of her post made on 5 June 2011:

> But I am complex, I am many things; I am an Arab, I am Syrian, I am a woman, I am queer, I am Muslim, I am binational, I am tall, I am too thin; my sect is Sunni, my clan is Omari, my tribe is Quraysh, my city is Damascus . . . And I am also a Virginian. I was born on an afternoon in a hospital in sight of where Woodrow Wilson entered the world, where streets are named for country stars. (Araf, 2011)

The following day someone who identified themselves on her blog as her cousin claimed Araf had disappeared and was feared abducted. Araf had made earlier posts worrying that she was being followed. Some frantic efforts were made to find her with a 'Free Amina Araf' Facebook group gaining 15,000 members. As this panic erupted, it was revealed that Amina Araf was more complex than she had admitted and was in fact a male, heterosexual, American living in Scotland called Tom McMaster, who saw the blog as a work of fiction he had been developing (Addley, 2011).

How was s/he found out? Some of the first doubts emerged when pictures posted by Amina supposedly of herself and her family were shown to be of someone else and had been copied from a Facebook page. Further attempts to contact Amina led to some false claims by her, particularly a request for a Skype session was refused because Amina claimed Syria was blocking Skype, even though only downloading of Skype and not its use was blocked. A face-to-face meeting was agreed but Amina failed to show up, claiming later that she had been followed and so aborted the meeting. Further, an earlier blog by Tom McMaster in which he used the Amina name was found and in it he explicitly stated the blog would be a mixture of fact and fiction (Guardian-MEL, 2011). The blog 'The Electronic Intifada' pursued this possibility of identity deception. They identified a married couple of Tom McMaster and Britta Froelicher as being connected to Amina in a couple of ways. They found that the only known address given for Amina Afar was for a property this couple owned. And they identified various pictures supposedly posted by Amina in Syria as having been taken at the same time and place as pictures posted online by Britta Froelicher (Abunimah, 2011). Within a short time, McMaster admitted his authorship and apologized, with the case gaining sensational and widespread media coverage.

Yet this is an old story, not in the specifics of Amina, sexuality and Syria which gave this particular example much of its power, but the old story of spoofing identity online. One term for it is 'sock puppet' and it has recurred in many ways all through the history of computer-mediated communication. Consider this second story from 1982.

The message boards on the CompuServe system in 1982 were accessed not over the internet but, mainly, using modems ringing up over phone lines. A woman named Julie appeared on them, she typed that she was a neuropsychologist who had suffered a devastating car crash leaving her mute, paraplegic and typing using a head-stick to touch buttons on the keyboard. Julie became a presence on the message boards, characterized by her commitment to helping others and drawing in others with her story of rebuilding confidence through finding her voice online. She began practicing online sex and told of her life offline which was transformed with a marriage and a resurrected career. At one point, the board heard, from her husband, that she was ill and near death and rallied to try and find her, though she stuck to her long-time position of refusing to meet face-to-face because of her disfigurement. Julie pulled through the illness, but when someone tried to contact her at the hospital she was supposed to have been at, they found Julie had never been there. From this point of suspicion, things then unravelled (Stone, 1995, pp. 65–81; Jordan, 1999, pp. 63–5).

Julie was, of course, not Julie but a male, able-bodied, psychologist called Sanford who had typed Julie into existence from the beginning. The illness and hovering near death was Sanford's attempt to deal with the tension his double-life was creating by killing Julie off, but he found the strength of feeling it aroused in those who knew Julie was too strong and he could not kill the virtual subject. Further efforts by him to resolve the situation, combined with the failure of Julie to be found in hospital, finally unravelled the situation entirely (Stone, 1995, pp. 76–8). Even in the 1980s, well before mass use of the internet, relations of style and identity were at work. Stone reflected on the Julie story:

> The hackers in my study population, the people who wrote the programs by means of which the nets exist, just smiled tiredly. A few sympathized with the women whom Julie had taken in, and understood that it takes time to realize, through experience, that social rules to not necessarily map across the interface between the physical and virtual worlds. But all of them had understood from the beginning that the nets presaged radical changes in social conventions (Stone, 1995, pp. 80)

Where Stone sees social conventions, I have explored communicative practices.

In both stories, it is the style that writers are able to construct in online spaces that identify them and make them into communicators able to send messages, whether of the life of a lesbian in Damascus or the rebirth of a disabled professional. And both stories of virtuality unravel at key points because a different communicative practice is invoked. In Amina's case, this was through confirmation that Amina's pictures were not of her and that her claimed physicality did not attach to her identity. In Julie's case, it was when the body failed to be present in hospital where the identity said it would be. The practices that stabilize a writer and reader by constructing a body dismantled the style the writers used to stabilize their online identity by revealing the clash between the two communicative practices that allow dyads like Amina/Tom and Julie/Sanford to exist in the first place.

However, I do not interpret these as examples of the authority of the living body to identify individuals. Stone shows us two aspects of this. First, she explores how Sanford was unable to be rid of Julie by transferring her friends to him. He conducted a campaign in which he created an account as himself and had Julie recommend insistently that this new Sanford person was a wonderful person. But the body behind Julie, that male body which remained unchanged sitting at the keyboard typing both Julie and Sanford, could not create the style of Julie. Stone recounted that Sanford

> then began trying to make friends with Julie's friends himself.
>
> He couldn't do it.
>
> Sanford simply didn't have the personality to make friends easily on-line. Where Julie was freewheeling and jazzy, Sanford was subdued and shy. Julie was a confirmed atheist, an articulate firebrand of rationality, while Sanford was a devout, conservative Jew. . . . Sanford's Sanford persona was defeated by his Julie persona. (Stone, 1995, p. 77)

Second, Stone followed the reaction to Julie's unveiling to find that some women, even some who were deeply shocked and offended, came to terms with this and missed the Julie they had known and began to seek out that style in Sanford. One woman stated:

> I've been trying to forget about the Julie thing. We didn't think it through properly in the first place, and many of the women took risks that they shouldn't have. But whether he's Julie or Sanford, man or woman, there's an inner person that must have been there all along. That's the person I really like. (Cited in Stone, 1995, p. 80)

And it would not be surprising if Amina began to be missed and her 'inner person' was distinguished from the controversy and the body of McMaster. These two stories dramatize the point that we live within two communicative practices and these are both interrelated and sometimes contradictory. Two different types of presence are created, using different material performatives and these can now be compared and interrelations suggested. Following this, the general results of the state of being in these two communicative practices will be examined in a state of normalized communicative anxiety.

Style and body

Communicative practices have been defined as the creation of presence through material performatives that result in the stabilization of senders, receivers and messages meaning that transmission can take place. This is an overarching claim, meaning that sets of communicative practices are the same at this abstract, general level but can be expected to operate in different ways in specific spatial and temporal contexts. The aim of this definition is to direct attention to the intermingling of technological and cultural factors which create the possibility of messages. This focus on the conditions for creating transmission has been explored in particular spaces and times that addressed communicative practices that are pre-internet and internet dependent. Nothing prevents several such practices existing at once and a definition of post-internet communication is existing simultaneously in these two communicative practices. A further step in this analysis is then to compare them and to see if this allows connections and contradictions to be understood.

Stability has often been invoked in this analysis as a quality that communicative practices must establish. Derrida's and Butler's use of iterability raised the problem of the simultaneous necessity and impossibility of repetition, because each repetition either returns exactly the same thing, making it not a repetition, or returns something different, even if subtly so, thus also making it not a repetition. Communicative practices produce this paradox intensely as their aim is to ensure transmission of messages which by its nature must involve repetitions as messages are transmitted between senders and receivers.

Repetition is core to presence in Levinas' understanding of care and capture, which makes presence a conversation and a hostage-taking both of which require the repetitions of interaction. Stability in communication is then dependent on repetition even given repetition's difficulties. Butler opened up the way to understand this when exploring the social logic of performativity, and a related suggestion comes from Derrida's claim that the deconstruction slogan 'there is nothing outside the text' might be better understood as 'there is nothing outside context' (Derrida, 1988, p. 136). Stability comes not from resolving the abstract contradiction between necessary repetitions and the impossibility of repetition, but from the social context which allows what is important in a repetition to be identified and hence implements certain limited and imperfect repetitions. Stability comes from social action, from what Butler called 'sedimentation,' or we might think of this as social action that becomes recognized as the same social actions by forgetting their inevitable differences.

The problem of repetition, which must be resolved for stability in communication to emerge, is then not solved theoretically but practically in the various technologies and techniques that are enrolled in a communicative practice. As a communicative practice is explored, a range of actions will be found that taken together create repetitions and hence stability in transmission. Communicative practices are subject to tension, always generating recognizable and limited repetitions in the face of the failure of repetitions. Such practices as signatures, role-playing, greetings and farewells, use of emoticons and all the others found in the two case studies, function together to construct repetitions that are stable. Each of the two discursive practices has a central stabilizing dynamic which is generated from these various performances. Comparing these core forms of stabilization allows an exploration of differences and similarities in their practices on this central dynamic of any communicative practice.

A central dynamic for pre-internet communicative practices is authorizing the sender of the message through performatives that create a body that has touched the letter, this body is then taken as a guarantor of the sender's identity. This creates a powerful point of stability that embeds in these practices a privilege for the sender over receiver because it is the sender's body that authorizes messages. This is created with various performatives that can be repeated even if each is a little different; just because no two signatures from the same person are exactly alike does not mean that a social performative cannot be practised widely enough to allow signatures to be repeated. Whereas once signatures had to be literally witnessed, with a human who saw the signing travelling with the signature to repeat to the reader that this signature had been seen to be attached to the message, by the early nineteenth century, the practice of signatures was socially sedimented enough to allow repetition. Similarly, the style of handwriting and content such as characteristic concerns and types of stories, each connect to and reinforce each other, creating a dynamic set of everyday actions that create a body that guarantees the reader's identity as having produced the message. We can see this social logic of repetition also from when it is absent. For example, some things that might be familiar to mid-twentieth-century letter writers were not yet sedimented in the early nineteenth century. Letter writers in the mid-twentieth century would most likely expect not just a greeting but a date and place of writing to be included at the beginning of the letter, but the latter two were not present in the early nineteenth century.

Internet-dependent communicative practices reverse the privileges of reading and writing, through styles that have to be read consistently for the writer to be stabilized as an identity. The use or not of emoticons, the use or not of grammar and full sentences, the characteristic topics of conversation (humour versus game dynamics, or some mix) are all ways in online games for readers to stabilize the identity of those sending messages to them. These have to be repetitions, though each repetition is a little different; just as N sometimes responds <grumbles> and sometimes <looks over suspiciously>, offering a difference within what is also recognizably a repetition of N and Llirk's mocking greetings. The stability here again derives from social actions that refuse to recognize all differences but recognize some similarities as constructing a repetition. Again this is not a philosophical position, when in everyday action the philosophical problem (which remains) is simply transgressed by social actions that in turn generate stability and a working form of repetition.

In stabilizing the repetition of senders and receivers of messages, key differences emerge between the two sets of communicative practices under analysis. In particular, in pre-internet practices it is writers' identity that is offered to readers as the authority that created and sent the message, whereas internet-dependent practices require readers to recognize a style in the sending and receiving of messages. Senders must then perform in different ways in these different contexts, able to assume an identity that is translated into meaning in messages in pre-internet practices while having to maintain an unstable performance in internet-dependent. Operating with both practices thus means being able to shift between or integrate quite different subjectivities as a writer/sender and as a reader/receiver. In one context, a writer/sender has to establish their subjectivity by using various performatives that create a body that authorizes and then, in a different communicative context, must develop a subjectivity from the ways they are read using the different materialities inherent in internet technologies. Not only are communicative subjectivities quite different but a second difference emerges as the very matter of communication shifts as communicative practices shift.

The materialities that are relied on in the two communicative practices make a useful comparison, as though both require materialities this manifests in different ways. Materiality in communication is also a performative within a performative. The performative of matter puts certain things beyond the social, providing a basis in things that can be assumed and on which other performatives may build. Occasionally this comes to the fore, for example when the matter of ink for letters dissolves into Meyrick trying to invent different inks. This account of matter also means matter is not subject to one divide in which something is natural or social, but to various forms of action in which some things come to be more or less within the social. We can, by tracing the specific performances in which matter is enrolled and used, also trace the various gradations of being-beyond-social action; these will stretch from institutions that are taken for granted but once questioned intuitively appear to be social all the way to things that appear to be nothing but a thing. For example, while a regularly functioning postal service might be assumed (e.g. it is assumed seals are not randomly broken and letters read), it is intuitively an institution, as the Port Phillip Association demonstrates when the postal service is absent. If we move along then to what seems more material, Meyrick's ink might be considered a 'matter' in its neutrality when used, but this comes undone when we see Meyrick having to make ink and worrying whether it will work and carry his words. At a limit, we can see what might be considered entirely natural elements like the distance between England and Australia or the trees that need to be cut for wood to make ships to travel the distance. All such forms of matter, however, only make sense in the context of communication when they are part of constructing forms of presence and so are performatives enrolled by other performatives. For example, the distance between England and Australia which is such a powerful natural presence in colonial letter writing, determining such things as the time messages take and hence the constant accounting of letters sent and received, is in internet-dependent communication barely significant. In voice or video communication over the internet, the previously naturalized distance between England and Australia recurs only as a faint memory, perhaps when high-traffic levels over this distance produce lag or slow responses.

An example will illustrate the place of materialization in communication, from which we can then consider broad differences or similarities between pre-internet and internet-dependent communicative practices. Computers and the internet use and depend on electricity and this is rarely commented on except when considering environmental consequences of the electricity industry and the pressure widespread computerization produces for more electricity. In these discussions, the nature of electricity is not at issue, it is the effects of a particular organization of the distribution and consumption of electricity. We might therefore consider electricity as matter, an asocial thing on which internet-dependent communication depends. Yet electricity is not a thing independent of its production and distribution, electricity in the form of moving charged particles (often electrons but not necessarily) is one type of power that, once organized through wires and cables and choices between ac and dc and so on, becomes available to plug into a computer to power it. What is taken for granted as the thing electricity is in fact organized in ways that are intensely social for electricity as matter to be relevant to internet-dependent communication. Electricity is, further, only created as something beyond the social when it is defined scientifically, that is by humans through various social and cultural norms and institutions. Even the charged moving electron can be unpacked from its natural status, if we follow its discovery and definition through the socially and culturally charged collective actions that brought, and continue to bring, electrons into use (Barnes et al., 1996). If matter is the sedimentation of particular performatives that imply certain things are outside of the social and so can affect but cannot be affected by the social, then we can examine the characteristic materialities of particular communicative practices.

The materialities of pre-internet communicative practices need to maintain the message and its attached body across the space between sender and receiver. It must then rely on materialities of stability that can be transported and remain unchanged. Such materialities must struggle against fluidity or instability; the ink must not run, the seal needs to close the letter and retain the shape imprinted on it, the paper needs to retain is form. The materialities of internet-dependent communicative practices need to ensure movement and continuous presentation and re-presentation, so that styles are engaged in enough interaction that they can stabilize the receivers, senders and messages of transmission. Materialities fundamental to internet-dependent communication assume the possibility that identity markers are untrustworthy and that the content of messages can be played with by each intervening hop, yet they maintain the flows through which enough interaction will allow styles to gain purchase. An online post can be edited and changed, an email can be forwarded or replied to and content shifted, multiple identity markers can be generated and used; all these reflect that internet-dependent communicative practices rely on materials that flow.

Pre-internet materialities can be read over one hundred years later with handwriting sometimes feeling as bright as if only recently marked, and even if some elements go missing, such as wax seals which were often absent (though leaving a faint mark on paper where they had been), then this merely tells us that this letter has been read. The paper and ink bring the communicative bodies of Meyrick, Black and others to the reader, even today. Emails and forum posts are far more ephemeral, perhaps being edited, archived or deleted, and they convey far less immediately who was in communication unless a series of such electronic messages can be read. The sure knowledge that the shapes the fingers make with ink on paper will be the same shapes viewed by the receiver contrasts strongly with the characterless uniformity of typed individual letters that may be cut, copied, interpolated and re-presented with each transmission. Quite different materialities are invoked in producing these contrasting situations and are necessary to these communications.

Not only do communicative practices draw on existing materialities like wood or electricity but they also continue to affirm their status as materialities, using them as if they were socially neutral and able to be simply drawn on. The key form of performativity here is in assuming that a matter such as electricity is socially neutral and in making that assumption in communicative actions producing that very social neutrality. While pre-internet and internet-dependent materialities share this performative drawing on and building of materialities, as just noted they differ in the kinds of materialities they draw on.

Internet materialities transform into flows of indistinct matter, into the same 'stuff,' from which are reconstructed the interchangeable elements that will present a style. It might seem paper and inks are materialities that flow but each when used must be distinct to the body that writes, they flow only to become unchangeable. In contrast, internet materialities need to support a constant stabilization and destabilization. If we think of individual letters, such as the English a, b, c and so on, then these were produced by the matter of ink and nib in colonial Australia to take the unique form of the writer's handwriting. The specific shape and characteristics of a writer's letters helped other practices to produce the body that is the sender of messages. In contrast, in the chat of online games, these same letters are rendered entirely uniform and also malleable in being easily reinserted and moved around. This simultaneous sameness of shape and ease of use is provided by the materialities of silicon, electricity and so on, which must be repetitively used to construct a style which underpins transmission.

It cannot be the present concern to pursue electricity and ink further, but even the analysis so far notes how different materialities mark the different communicative forms and in turn reproduce certain things as matter. A few recurring characteristics of these different materialities shape and are shaped by their communicative practices. Perhaps, the most striking is the way the internet is often considered virtual and immaterial. This is literally wrong, as every transmission carried by the internet is carried in particles, stored in electricity, located in particular places and times, but it reflects the sense that these are highly malleable materialities, allowing moments of stability only on the presumption that they can turn into further changes. The materialities of internet-dependent communicative practices rely on their ability to change and to differ, whereas in pre-internet communication transmission requires a stability that stops and remains the same. Both practices engage materialities, but each organizes, produces and depends on materialities of different kinds.

Two substantial differences between the communicative practices under analysis can now be summarized. On the stabilization of transmission through definitions of senders, receivers and messages, the two communicative practices differ in ways that seem opposite, with one emphasizing senders and the other receivers. On the materialities these two communicative practices rely on and also reproduce, they also draw, on the one hand, on things that can be coordinated to produce stable objects and, on the other hand, on things whose nature is malleable. Across both however remains the core communicative moment, the lived or average everydayness of communication in transmission. Whether it is a typed greeting in guild chat in a virtual world or the greeting and farewell that defines a colonial letter's content, in both communicative practices the moments of transmission remain. This strongly supports the similarity between these two communicative practices being the creation of conditions which ensure transmission can be stably repeated. These conditions were argued to be the creation of presence through material performatives. While we can see distinct differences in the nature of presence and material performatives between the two communicative practices, we still see in both presence and material performatives.

This suggests two things. One is that a theory of communicative practices as the creation of transmission through presence produced by material performatives has had insight into the fact that messaging is possible. The connection, or collision, between communication studies and cultural studies seems to have produced, particularly from Milne's insight into the importance of presence when comparing different forms of communication, a theory that grasps communication before and after the internet.

The second implication is that the two communicative practices engage simultaneously in everyday actions of transmission. Both communicative practices produce ways of stabilizing a sender, a receiver, a message and means of messaging. The ABC of communication lies in the sending of A to B by C.

The experience of transmission involves precisely that of a repetition whose conditions are in a sense unconscious, similar to the way de Certau defines everyday conversations as 'verbal productions in which the interlacing of speaking positions weaves an oral fabric without individual owners, creations of a communication that belongs to no one' (de Certau, 1984, p. xxii). The unconscious of transmission is the presence that communicative practices establish and rely on. And, precisely because presence drifts into this social unconscious, these practices that no one owns, transmission will seem to those engaged in it to be simply the sending of messages because the conditions for such sending can be forgotten.

Transmission is then not only the result and product of communicative presence but it is also a zone of contradiction in which is lived the assumption that no matter the communicative practices and how different or how many of these there are, the sending of messages can be lived as one process. In this zone, the different roles of bodies, styles, materialities of stability or malleability are diminished in the acts of transmission which are both made possible by all these and forget that there are different bodies, styles and materialities that are necessary. Transmission can be lived as one, even though in the twenty-first century it is created by (at least) two extensive, complex and contradictory communicative practices.

What is the result of this zone of communicative contradiction? What social and cultural effects are there from an everydayness that is communicatively incoherent because of contradictions between communicative practices but which somehow remains coherent in the unifying moments of transmission? Here we meet a key broad consequence in communicative anxiety, whose examination will allow the theory of communication and the internet to offer some ideas about the cultural and social effects of existing simultaneously with pre-internet and internet-dependent communicative practices.

Anxiety

Mobile phones combine in one technological artefact both pre-internet and internet-dependent communicative practices and because of this they provide an example of what living in both communicative practices means. On a mobile phone, the voice can be heard which allows the body and identity to be stabilized in similar ways as signatures can stabilize. A second common point is that there are greetings and farewells that delimit the message. For example, Caron and Caronia's investigation of teenage mobile-phone use found that despite the phone listing who is calling, thus the receiver and sender both know who the other is, both sides still employ a greeting and farewell routine. Like the greeting and farewell in the letter, this confirms who is who, in this case authorizing by the voice, and it marks out where the meaning in the message begins and ends. They identify this everyday, unconscious and repeated practice by finding the exception in that lovers do not always employ a similar greeting and farewell but sometimes simply pick up 'a conversation on a communication channel that is already established and permanently open' (Caron and Caronia, 2007, p. 154). Voice telephony is in these ways consonant, at the level of communicative practice, with practices found in early nineteenth-century letters. At the same time, mobile phones offer text messaging, a form more consonant with internet-dependent practices, and often internet connectivity ensuring connections to email and social-media sites that rely on internet-dependent communicative practices. In one materialized complex form, the mobile phone takes us to the heart of what it means to live simultaneously in two communicative practices.

One of the recurrent controversies around mobile phones is situated at this intersection in the often commented on, both in academic and popular analyses, phenomenon of mobile phones breaking, threatening or changing co-present interaction. The attention given by those who are physically face-to-face to each other is now never free from those who can be reached or who might be reached through mobile phones, what Caron and Caronia call the 'ghost participants' (Caron and Caronia, 2007, p. 4; Ling and Donner, 2009, pp. 107–12). In Levinas' sense, mobile-phone users are hostage to all those who might be able to connect to them. The controversy often revolves around mobile-phone users prioritizing communication through the phone over communication to those physically co-present. The everydayness of this controversy in manners and social interaction was marked in 2010 by an advertising campaign for the new Microsoft Windows Phone 7 operating system, whose tagline was 'It's time for a phone, to save us from our phones.' Briefly exploring this advertising campaign will help clarify this point.

The television advertisement begins with intercut shots of people using their phone in everyday situations – while drinking a cup of coffee, sitting underneath some candle-lanterns, at the beach – and then gradually gains in intensity with shorter, swifter intercuts of scenes and rising intensity of music, while at the same time the context in which phones are used becomes more absurd. First, someone is riding a bike on a busy pavement while using their phone, then someone is in the shower holding the phone just outside the shower curtain, and then there is someone jogging along a busy pavement while using both hands to thumb-type with phone held in front of their face. About midway through the ad, these scenes reach someone being massaged with the masseur using their elbows on the patient while they type on their phone; the quizzical patient looks up and utters the defining word of the ad 'Really?!' The ad progresses from here with increasingly odd scenes provoked by mobile-phone usage, nearly always commented on with an angry, quizzical or astonished exclamation of 'Really?': one man standing peeing at a urinal drops his phone into the urinal and immediately grabs it back; a man in bed ignores an attractive women in lingerie gesturing toward herself; a father sits on a see-saw focused on his phone while a bored child sits stuck up in the air on the other end; someone with snorkelling gear on sits with their feet in the water off the end of a boat focused on their phone as a large shark approaches; a bride comes down the aisle tapping her phone; a surgeon stands above a patient already cut open and gazes at his phone and a father playing baseball holds up his mitt to his child to catch and looks not at the child but at his phone. In each case an aggrieved or watching party offers up the 'Really?!' and looks on in puzzlement; in the baseball case, the child throws a ball violently at the father hitting him on the head. Right at the end, we are exhorted to buy a phone using Windows Mobile Phone 7, so that it can save us from 'our phones'. The final shot is of two young people sitting down to dinner in a restaurant, no phones in sight, while the caption 'Be here now.' appears between them (ironically in an ad for mobile phones suggesting that life is better without mobile phones).

While it would be foolish to mistake such advertisements for a re-presentation of reality, such advertising campaigns need a resonance with wider cultural norms and mores to succeed. The 'Really?!' must also be resonant enough to overcome the paradox in which this is an advertisement promoting a particular kind of phone while at the same time suggesting mobile phones are ruining everyone's lives. Such interruptions of the face-to-face by the mobile are also present in research on mobile-phone usage; Ling and Donner call it the 'disruption of co-present interaction' (Ling and Donner, 2009, p. 107) whereas Caron and Caronia note:

> When two people keep their mobile phones on when meeting for lunch, one must ask what has become of a simple conversation. The rights and duties, expectations, and even face-to-face manners of participants must now be negotiated with those of *ghost participants* who may join interactants and enter into the occurring conversation. People living 'out there' can be 'here' at any time, thus changing the participation structure of the event. (Caron and Caronia, 2007, p. 4)

Most analysis progresses from this point assuming that this clash produces problems and discussing the ethics of the face-to-face world having to reorder its priorities because there are now many who are not physically face-to-face but who are co-present through the mobile phone. However, it is not clear why such a clash produces the anxiety and uncertainty that both research and popular indexes like the Really?! ads find. This can particularly be seen in relation to Caron and Caronia's tracing of how young people's use of mobile communication involves a reconstruction of what being polite means. They emphasize that young people create a new understanding of politeness among themselves related to the novel situations mobile phones create.

> Faced with the theatrical performance embodied in all public uses of mobile phones, young people reflect on the difficult and always precarious equilibrium between individual and collective concerns. Even though they explore this problem in a perfectly circumstantial manner, they nonetheless weigh difficult issues such as respect for others, the need for shared rules in social life . . ., the possibility of violating rules . . ., and the inevitability of punishment. (Caron and Caronia, 2007, p. 228)

Similarly, Licoppe's concept of connected presence – 'in which the (physically) absent party renders himself or herself present by multiplying mediated communication gestures up to the point where co-present interactions and mediated communication seem woven in a seamless web' (Licoppe, 2005, pp. 135–6) – suggests that the ethics of everyday communication will change around mobile communication but suggests no immediate barrier to it also being calm rather than anxious. It is not impossible to imagine ways of stabilizing and moving on from the supposed disruption of mobile communication. For example, many will be familiar with a moment in which someone we are with receives a mobile-phone call and whispers to us, 'I'll have to take this, sorry,' and then departs from the embodied world into the mobile world. If all that was happening in such situations as this or those referenced by the Really?! ads was the emergence of a new ethics, then such moments would not have significant cultural resonance because they would simply be a record of people behaving badly, not a common and anxiety-provoking situation. If all that were going on was the adjustment of the ethics of the everyday, as Caron and Caronia, Ling and Donner and Licoppe suggest is possible, then it is not clear why this situation is a recurring and awkward one rather than simply a process of readjusting social ethics that we will all become used to as normal. Instead, it seems that the intersection in the mobile phone of those who are present and those who are an absent presence, indexes not only a socially awkward moment when these contradict each other but something in addition that results in anxiety and difficulty in that moment.

Is there something in this coincidence in the mobile phone of pre-internet and internet-dependent communicative practices that provokes something like anxiety and unease? Is there something that explains the cultural resonance of Really?! and the compelling way people currently pass through life constantly referring to their mobile phone (working in a busy inner-city university, it seems to me I cannot walk for more than a few minutes without having to avoid someone who has their eyes fixed on their phone)? What seems to be indexed in these interruptions of physical co-presence is an anxiety and displacement in which people are unable to situate themselves in the one place. In all the short scenario's the Really?! ad uses, there is this same structure of someone facing a compelling experience in one realm but being unable to disconnect from a contradictory realm. What can we make of what seems to be a kind of widespread and widely recognizable social anxiety?

Lacan's early concept of anxiety was that it reflected the potential fragmentation of the subject, which had been formed 'externally' through confrontation with the Other. Lacan, famously, located this moment of formation in the mirror-stage of development, equally famously Lacan complicated the concept of the mirror-stage throughout his career and developed a later concept of anxiety in relation to his concept of the real (Lacan, 1977a, pp. 5–6; 1977b, p. 41; Redman, 2000, p. 11). While there is no intention (or need) here to follow Lacan's complex and large body of thought, his suggestion that anxiety relates to threats to the construction of Self in relation to the Other invokes the notion of presence and opens up a turn back to theories already discussed which can be developed to understand communicative anxiety.

Earlier Heidegger's concept of anxiety was touched upon because it occurs in the idea that average everydayness is a drowning of 'being-with' in the concerns of the masses. Heidegger then casts anxiety as the state brought about when someone drops from their average everydayness and considers the possibilities of Being in a primordial sense that questions existence. Accordingly, anxiety (or *angst* in the German, or dread or unease as it may be variously translated) is a questioning and uncertainty in which a being shifts or is caught between Dasein and the revelation of Being and the average everydayness lived from moment to moment which conceals Being (Heidegger, 1980, pp. 225–7; Polt, 1999, pp. 76–80). While Lacan and Heidegger are arguing from different theoretical and historical backgrounds,[2] they are similar in discussing anxiety as a moment in which a subjectivity's settled, stable or still state is disturbed with that disturbance raising potentially fundamental issues of self and being for a subject.

This connects to the sense of communicative presence, drawn not only from Heidegger but also from Levinas, Derrida and Milne, in which presence is the construction of the face-to-face, in the responsibility, care and inescapability of others and of the Other. Presence in communicative practices has to be created and maintained for transmission to happen, and this presence involves constructing communicative subjectivities for senders and receivers of messages through various materialities and performatives. When these subjectivities are different because of different communicative presences and they start to collide, then perhaps they produce anxiety in their unpicking of each other. Being-with/becoming-with as care and capture, that is communicative presence as the generation of being-with through the becoming-with and maintaining of senders and receivers, is threatened when different forms of such subjectivities are simultaneously at play. This undermining can be considered anxiety in which a certain subjectivity is threatened. What is produced is an endemic state of communicative anxiety that mobile phones most prominently provoke because they carry in one artefact both communicative practices.

This is not a matter of assuming a need for a unified subjectivity, the anxiety provoked here is not the result of there being several subjectivities which cannot be coalesced into a whole. The arguments about decentred identities and subjectivities which have been extensively explored in the last 30 years are not being ignored or denied here (du Gay et al., 2000). Rather, the issue is more specific than a general definition of identity because it relates to two communicative subjectivities that perform the same role in transmission as lived in everyday moments but which achieve such moments through contradictory forms. It is not anxiety from being decentred, but anxiety from contradictions that dismantle each other's practices; it is anxiety from creating an identity that can be part of transmission that is then disrupted by a different identity struggling to coalesce into its form of transmission. Two aspects of this are illustrative; the prominence of readers or writers and the swift pace of shifting between practices.

As discussed already, there is a difference in emphasis between pre-internet and internet-dependent practices in which the former offers a priority to the sender of messages constructed through a body/subjectivity identification and the latter depends on a priority from receivers who legitimate senders based on styles of messages and sending of messages. Communicative practices generate and maintain these different kinds of subjectivities through various material performatives. As discussed the 'strong bindings' found in nineteenth-century letters worked to create a kind of body and subjectivity that could be identified with each other and so authorize a writer whose meaning could then be read. If two such subjectivities, each generated with a similar purpose because each is part of a presence that enables transmission, are in fact different to each other because they are engaged in different technologies, materialities and performatives in a different kind of communicative practice, then the everyday includes these clashing and this clash will occur in the moments of transmission.

As moments of transmission come and go, moments only made possible by all the machinery of the communicative practice, then subjectivities must be shifted, bodies change their meaning, authorization moves from the writer to the readers and so on. As each text is sent, each phone call is answered, each email is written, then various subject positions are invoked, used and dismantled. In both pre-internet and internet-dependent communicative practices, transmission, with its B sends to C message A structure, is created but what A, B and C mean are different and are maintained through different performatives. Such differences are material with internet-dependent practices having the flow-like materialities of electricity while pre-internet rely on stability and sameness.

The anxiety described here is that of a constant dismantling and recreation of different elements of transmission. The moment when someone's mobile phone beeps in the middle of a conversation and while still talking someone takes out their phone and their eyes move from the body next to them to the screen and the email message on it and, while still nodding and seeming to engage in the face-to-face conversation, then starts to tap out a reply; this moment is one of intense anxiety as the different subjectivities that will allow both kinds of transmission clash. This moment, an everyday moment, involves differences in what it means to be a reader, writer, message and means of messaging for all the participants in the moment. This clashing results from the nature of the two communicative practices because they invert each other and produce transmission in different ways.

While research suggests that many people find it irritating to have their face-to-face interaction trumped by a mobile-phone call, these results tend to focus on the fact of interruption and the increased chances of such interruptions with mobile-phone use (Caron and Caronia, 2007, pp. 224–31; Ling and Donner, 2009, pp. 109–11). The anxiety I am laying out is different because it invokes two types of communicative practice, one of which revolves, in part, around the body – and physical co-presence and voice can both generate and draw on such bodies – and the other around readers' reading a style. One way of seeing this is to consider those who might try to escape this anxiety by refusing one of these communicative practices, for example in obsessive online behaviour involving excessive time spent game-playing or hacking or the related online phenomena of otaku. Psychological takes on internet-dependent practices, as well as some moral panics, have often focused on the idea of 'internet addiction'. Such a phenomenon we might now understand as attempting, at least in part, to refuse pre-internet communicative practices so that someone can live their everyday entirely and solely generating transmission through internet-dependent practices. While it would be reductive to suggest communicative anxiety is all that is going on in such cases, it is also legitimate to point out that such disavowals of certain communicative practices are part of obsessions and addictions (Joinson, 2003, pp. 55–9). Communicative anxiety comes not from one voice interrupting another, this is something handled within shifts in collective understandings of politeness and related social mores, but from a different sense in which what a voice, body, or message means is both invoked to create transmission but is also destabilized and undermined by a different form of transmission. Communicative identity after the internet imports a necessary anxiety into the swirling whirlpool of decentred identities of the twenty-first century (Touraine, 1995, pp. 204–31; Melucci, 1996, pp. 28–33; Turkle, 1996).

But is such anxiety a passing phase? Is it a specific historical moment linked to the introduction of a new communicative practice that will pass as communicative practices are adjusted and different social interactions become routine? The spread of such instances of different communication as 'text-speak', such that omg and lol have both been added to the Oxford English Dictionary, reminds us that what are at stake are not fixed universals but the particular social and cultural practices that allow transmission to reliably and repeatedly occur. As such, there is no question that such practices can and do change.

However, the point here is not that practices are immovably fixed, static forever, but the kinds of dynamics that are becoming embedded. The answer at this point is surely not that more and more people are becoming familiar with and are using simultaneously forms of transmission derived from, at points, contradictory communicative practices but what does this mean? What is being embedded as 'normal' within our everyday moments of sending and receiving messages? This analysis argues that anxiety provoked by shifting between different forms of transmission is precisely what is becoming gradually normalized and that while we remain dependent on using two kinds of communicative practices, we will also produce a form of communicative anxiety in which subjectivities are contradictorily lived and that this can itself be considered a normal everyday state.

Care of the communicating self

Identifying and exploring the idea of a communication-based anxiety which undermines an aspect of post-internet subjectivity connects communicative practices to wider social and cultural practices that intersect, enfold and are engendered by communication. It becomes difficult to consider the anxiety provoked by communication without starting to wonder how that affects or is affected by the 'care of the self' in the twenty-first century. Such a topic provokes consideration of factors well beyond the scope of the evidence and theories presented so far because the connection is to the many other aspects of society and culture that involve communication. Instead of extending these arguments too far, the final point to be made is then simply that there are these connections and that from this analysis of communication several key threads offer ways to begin considering the affects being after the internet is having.

First, an opposition of body and style has recurred throughout this comparative analysis and has become a key figure for considering communication in its wider contexts. As analysed earlier when considering Butler and materiality, there has been attention paid to the body by cultural and social theory. This attention has re-appropriated the body as a cultural factor and particularly politicized it by examining the ways the body is gendered, raced and so on. Analysis of communicative practices both draws on this extensive work but also suggests back that a further factor in understanding bodies is the communicative body and its generation, or not, within communicative practices.

The body appears in pre-internet communicative practices as a fragmented, constructed entity that is defined by its ability to authorize and stabilize communication. The body in these communicative practices leaves its mark in numerous ways, each of which indexes and creates the authority that flows from readers being assured that a particular body has touched a particular communication. This body is not whole, though it imagines a whole body as the author of the message, because it only exists through its various techniques; signature, seal, writing style and so on. One, perhaps neglected, element to understanding the fragmentation and analysis of the body, breaking it down into multiple elements such as how it becomes gendered, raced, sexualized and so on, is the common experience of this fragmentation in authorizing transmission as coming from a particular subjectivity.

This authorization is not restricted to the written letter. For example, this chapter has touched on mobile phones and suggested that these combine pre-internet and internet-dependent communicative practices, because the voice is carried by the phone and this provides it with the bodily anchor characteristic of pre-internet communication. This voice is, however, fragmented from the rest of the body, as over the phone the body is carried by timbre, intonation, accent and other elements that might allow any of us to instantly recognize a familiar voice. The body is not here carried by itself as a simple whole that is self-present face-to-face but is fragmented into elements. The communicative body before the internet is fragmented and represented by those fragments, from which a unified, whole body is imagined that is identified with a subjectivity from which a transmission is sent.

While communicative bodies offer a further complication to analysis of the sexed, raced, gendered and more bodies of cultural theory, the styles of internet-dependent communication can be seen to contribute to the creation of such politicized terms as gender, race and sex. Styles, as we have seen, rely on the collective generation of the stability of transmission, through readers identifying certain styles with senders of messages. These styles, as already argued, draw on various repertoires of interactions that are understood and generated collectively. Styles rely on being able to draw on and, in turn, may alter existing repertoires of action. The history of internet communication is not free floating but instead demonstrates particular places, times and peoples as key to its creation and spread, and in these times and places, only some repertoires have been generated embodying particular cultural ethics and politics.

In particular, and as often commented on, the gendered style of online communication with its characteristic aggression and mocking, as seen in such things as flame wars and trolling, derives from a number of sources. The gendered nature of online discourse has been discussed and argued about for a considerable amount of time (Dery, 1994; Cherny and Weise, 1996; Donath, 1998). More recent work has established strong connections between early personal computer and internet development, in particular the dominance of networking as a paradigm for online communication, and the US-based 1960s counterculture. The latter has to be situated within US-state funding for communications development, primarily from military-research funding. Even within a sector heavily dependent on state-military funding, a discourse that emphasized a competitive individuality relying also on an all-too-easy resort to linguistic aggression was important to cyberdiscourse (Markoff, 2005; Turner, 2006). As Turner's study of the US counterculture and discourses of cyberculture argues there were

> those who promoted the Internet and the network mode of production as evidence of a new stage in human evolution. Like the communards of the 1960s, the techno-utopians of the 1990s denied their dependence on any but themselves. At the same time, they developed a way of thinking and talking about digital technologies from within which it was almost impossible to challenge their own elite status. On the communes of the 1960s, the rhetoric of consciousness and community contained little in the way of language with which to describe, let alone confront, a less-than-egalitarian distribution of resources. The same was true of information theory and the universal rhetoric of cybernetics. In both cases, human power was an individual possession, born of the proper use of technologies for the amplification of awareness through access to information. (Turner, 2006, pp. 260–1)

Turner's detailed and powerful historical account is of only one discourse, albeit an important one, which was part of founding cybercultures. Such cybercultures will be important collective resources from which communication styles draw.

This was seen earlier in the resources for communication that gaming offers. As Turner's work demonstrates the embedding of certain libertarian assumptions in cyberdiscourses can be related in various ways to the masculine assumptions those analysing early cybercultures found (Cherny and Weise, 1996). This libertarian and masculine discourse remains important to repertoires available for internet-dependent communication, as Malaby's analysis of the creation of the persistent virtual-world Second Life demonstrates (Malaby, 2009). Other connections from the history of cybercultures are also possible, such as the early debates over the importance of ASCII as a character set for making English the presumed language of internet communication, which also connects to the initial international digital divide in access to the internet. There are also other contributors to cyberdiscourses, for example the discourse of Usenet or of the 'grassroots internet' Fidonet (Jordan, 1999, pp. 37–46).

It is not a matter of identifying all such histories and discourses here, but of noting the deep connection between styles as a key element of internet-dependent communication practices and the historical contexts of the emergence of this communicative practice. In this context, we find various elements which, because styles depend on collective repertoires of action, become embedded in these communicative practices. The politics of such communication, that is the personal, everyday politics of transmitting and receiving messages, is in this way joined to the historical politics of the emergence of the internet and cybercultures.

Where styles allow an opening to understanding the ways that the politics of culture and society have become embedded in internet-dependent communicative practices, the body offers ways of moving out from pre-internet communication and connecting to the politics of culture and society. In the latter, an additional understanding of bodies emerges in the ways fragments refer to a whole body that authorizes transmission but only fragments of the body are ever in action. In the former, the politics of the internet's creation, development and maintenance are embedded within the things that make communication possible when using internet technologies. Of course, in neither case does this mean that things may not or will not change. Instead, it is a matter of identifying where change may start from and then to begin to assess what change may be necessary for different types of futures.

Really?!

Anxiety, fragmented bodies and cyberdiscourses have been explored looking at how communication after the rise to mass use of internet technologies has become a state of existing within two contradictory communicative practices. This produces both existential difficulties, in the mutual undermining of communicative subjectivities, and political projects, in the identification of a communicative body and of the integration of the politics of cybercultures into communication.

The power and importance of communicative anxiety may well be the most significant factor here, not because it is existential but because of its potential for becoming normal. One of the difficulties in looking at communicative practices is looking at things that seem obvious and simple but yet have been shown to be important; that a signature indexes a body might seem obvious, yet it is only by paying attention to the seemingly obvious that the ways transmission is created can be identified. In short, that something is obvious does not make it unimportant. In the same way, the normalization of communicative anxiety may be becoming obvious and taken for granted because people now live with two concurrent communicative practices. As people become increasingly fluent in internet-dependent communication, and this is surely becoming widespread in many parts of the world particularly through the popularity of social media and with the breakthrough into popular imagination and mass use of online games and the spread of 'smart' phones, then for more and more people it becomes normal to learn and use two different communicative practices.

Normalization however here does not mean the elimination of anxiety, as communicative subjectivities remain in conflict across moments of transmission. This is the difference between normally meaning resolved and normally meaning usually experienced. Communication after the internet involves clashes between subjectivities that may disturb, may offer conditions for the pathological reaction of refusal of one communicative practice and will need to be coped with on a daily basis. Anxiety in communication has become widespread.

'Really!?' indexes this widespread understanding of the anxiety provoked by having to maintain subjectivities that are generated differently but address the same repeated moment of transmission of meaning. Anxiety is a constituent part of the personality of twenty-first-century communication.

Notes

1 Lee, 2001.

2 The relationship between Lacan's thinking and Heidegger is too complex and extensive to go into here. See Rockmore, 1995, 26–7.

6

Signature: Flow and Object

In Oxford undergraduate days, now alas some fifty years ago, a very great friend – the late Lord Willoughby de Broke – used to declare that the best time to live was during the period after the Battle of Waterloo.

The early chapters of this book are laid in that period, and what a contrast to the conditions under which we live now! Wireless, telephones, motor cars, and aeroplanes were not even thought of, and even the steam train was not in operation.

Sir William R. Campion[1]

Communication change

Sir William R. Campion reminisces about the age after the final defeat of Napoleon and reminds those of us familiar with the aeroplane, and to whom the steam engine is now a novelty, hobby or a memory, that change depends on where you stand. Changes in communication are the same and these arguments about communicative practices have attempted to trace changes in communication viewed from the early twenty-first century. I have tried to draw on what was, in 2012, already over 50 years of computer-mediated communication and will soon be 50 years of the internet,[2] to identify a broad shift in communication that places all of us somewhere after the rise of the internet, with its characteristic technologies, cultures and norms now embedded in our lives.

Social change may occur in a long or short time, for example we can compare the 'nearly fifty' years of the internet to Campion's hundred year melancholy for the post-Waterloo era. We can also compare these to a social change from over a thousand years ago that took three hundred years in Clanchy's argument about the shift from memory to written record. Clanchy's work is interesting here because it could be interpreted as an analysis of a shift in communicative practices, from the time when memory and spoken word were overtaken by writing in a way that distanced the bodies of senders and receivers from their messages. Clanchy (1993) does not undertake his analysis from a communication standpoint, nonetheless many of his claims provide an interesting counterpoint to arguments so far.

One example from Clanchy makes this point. Clanchy notes that at the beginning of the three hundred years he surveys, in Edward the Confessor's reign that ended in 1066, only the king possessed a seal or signatum, whereas by the end of Edward First's reign in 1308, everyone, even serfs, were required by law to have a seal used to authenticate documents (Clanchy, 1993, p. 2). This is a major shift, involving the generation of technologies of communication that ensure messages can be authorized and travel without bodies being co-present. This, of course, was neither a sudden nor an even shift. For example, in 1216, 'Louis, clerk of Rockingham . . . even varied the way he wrote his own name, as the idea of a distinctive signature was unfamiliar in England, where the authenticity of a document depended upon seals and the word of named witnesses' (Clanchy, 1993, p. 128). The role of the named witnesses and the seals appears as a transitional period as norms shift to create the consistent signature and seal and to trust both. The seal here appears as a midpoint between only trusting the co-present body and trusting the signature. What Clanchy in many ways traces is a transition from a fear of inauthentic communication because bodies were no longer co-present to a stabilized at-a-distance communicative practice. Socrates voiced a similar fear when he recounted the legend of the God Theuth who invented writing and presented it to his king, the God Thamus:

> when they came to letters, This, said Theuth, will make the Egyptians wiser and give them better memories; it is a specific both for the memory and for the wit. Thamus replied: O most ingenious Theuth, the parent or inventor

> of an art is not always the best judge of the utility or inutility of his own inventions to the users of them. And in this instance, you who are the father of letters, from a paternal love of your own children have been led to attribute to them a quality which they cannot have; for this discovery of yours will create forgetfulness in the learners' souls, because they will not use their memories; they will trust to the external written characters and not remember of themselves. (Plato, 1892)

Writing produces forgetfulness of the soul. This prioritization of speech is something Derrida, as has been traced earlier, identifies and argues against in his discussion of Husserl. In a sense, Derrida philosophically contests the consequences of centuries whose communicative practices were derived first from speech and then from ways of creating transmission on the model of speech, even when senders and receivers could not be co-present. By ever so briefly, and through Clanchy's eyes, looking back to the time when pre-internet communicative practices may well have been emerging from embodied communicative practices, we gain some perspective on what is being claimed about changes in communicative practices with the rise of internet technologies.

The hypothesis outlined in the first chapter foretold a certain story. This story has been traced through the abstractions of presence, materiality, practice and performativity. The story has been told in the registers of the letters of Meyrick and Black and the game-playing of Nyd and AS. With the story told, two complex ways of creating transmission, by different material performatives that generate presence, were identified. While in both transmission is clearly possible, with its characteristics of senders, receivers, self-identical messages and means of messaging, the kinds of transmission are different. From this it became possible to identify the anxiety that being after the internet embeds in communication, where subjectivities are undermined as everyday moments require invoking two different communicative practices.

Signature

Much has been said so far about the signature as an object, the tracing of lines to produce a name, moreover it is a repeated tracing that is a pre-eminent example of the problem Derrida argues inheres in repetition, that each signature is necessarily different but is also necessarily the same. But, so far, a second sense of signature has been underplayed, that in which signature references a characteristic way of being or doing. We can talk of the 'signature' sound a musician might have, or the signature move of a sporting star. In some cases, a particular innovation or stylish way of doing something can be so closely associated with someone that the move takes their name; for example, the Cruyff turn in soccer named for Johan Cruyff.

The signature thus has two aspects: it can be an object or thing which marks an identity or it can be actions or ways of being through which particular characteristics are displayed. These two interact through their allegiance to the same signifier; Cruyff could be said to have signed particular matches and there is a certain characteristic to Cruyff's handwriting that allows Cruyff's signature to be repeatable as a marker. This coincidence emphasizes the way anxiety is embedded deeply within the coincidence of signature as an identity marker and signature as a style. It also opens up the last points of this analysis of different communicative practices.

Having raised the dual meaning of the signature and having utilized so often in these arguments the nature of the signature, it might seem odd to have so far failed to examine digital signatures. Looking briefly at digital signatures will allow this story of communicative practices to come full circle and return to the idea explored at the beginning concerning metaphors between online and offline and their instructive failures. Those familiar with digital signatures may have several times wondered if my argument was at points simply missing the boat and had failed to take account of the hard identity marker of an encrypted signature in internet contexts. After all, there is no doubt about the huge growth in online shopping, which is dependent on such digital signatures swapped between online sites that authorize and stabilize such financial communications. The problem however remains that a digital signature is simply an encryption key, ultimately a set of numbers, and further practices are needed to ensure such a signature can be associated with an identity.

One useful example of this is the web of trust first invented with version two of Pretty Good Privacy, a program enabling public-key encryption. Public-key encryption allows someone to take the secret key that allows decoding of messages they have encrypted and to break that key in half. One key remains secret and private. The second key is linked mathematically to the private key but in such a way that the private key cannot be derived from this second key. The second key can then be made public. The process is then that sender A encodes a message with their private key. Receiver B can then use the public key to decode the message. In this way, A and B can pass encoded messages that no one else can read without having to also find a way to pass the secret code to unlock messages between them. Public-key cryptography was a revolution because it overcame one of the oldest problems of encryption in how to securely transmit the key to decode a message to the person reading it. This also creates a digital signature because B can be sure that the message is from A because only a message encrypted with A's secret key can be decoded with A's public key. This also opens up what seems like a hard marker of identity available on the internet in someone's public key (Levy, 2000, pp. 66–86).

However, even with a revolution in the concepts of encryption added to sophisticated mathematics that underpinned working versions of public-key cryptography, there remained the problem of internet-based identity markers. The public key is neither the person nor a guarantee that a person placed that key online or created that key. In short, how does B know that they have A's public key? Without that trust the whole scheme fails to work and the recurrent problem of constructing identities that can authorize communication recurs. A digital signature remains unlike a handwritten signature because it fails to be marked by a body. Attempts to ameliorate this problem confirm it. One of the innovations of Pretty Good Privacy version 2 was an attempt to deal with the problem by creating a 'web of trust'. In such a web individuals who wish to share public keys are able to testify to the legitimacy of someone else's key. For example, if identities A, B, C, D and E all have public keys and are certain of some other identity's keys then they can legitimize various members to create a web of trust. A may confirm that B's and E's keys are legitimate, while B may confirm C's and D's keys and E confirms A. But A and E confirm each other so E may trust A's confirmation of B, and so when E wants to send a message to B they can trust B's key because A confirms it as legitimate. Once B turns out to be trustworthy then C and D can enter the web of trust as they are confirmed by B (Levy, 2000, pp. 201–3). Such a web can extend such that a wide range of people have a reason to trust other identity's keys. If we probe this a bit more though we can ask why any of these identities should be accepted as legitimizing anyone else? The web of trust still does not solve the problem of authorization rather it distributes it.

There are two typical answers to constructing a successful web of trust by establishing authority for some identities to confirm other identities and these answers return us to different forms of communicative practice. One way of successfully confirming someone's key is to physically meet them and to exchange keys by exchanging some physical media, such as a data stick or, in the old days, a floppy disk. Pre-internet communicative practices for authorizing identity here reassert themselves at the heart of the digital era because the signature is here launched for use by confirmation through the co-presence of bodies. Though a common practice, it should also be noted that even the co-presence of bodies does not absolutely ensure an accurate key as someone could have transferred the wrong key or had a key unknowingly substituted; this is unlike the handwritten signature which connects the body and identity by someone being able, ultimately, to see the body produce the shapes that make up the signature. The second way of successfully confirming is through styles of interaction in which people assess from the ways they interact that they come to trust that a public key is from a particular identity. If someone turns out through their communication styles to be the person they claim to be then the public key becomes evermore solidly confirmed and can finally begin to confirm to others that this key should be trusted. Even within the complex mathematics and coding of encryption, the fundamental duality of the signature remains. In pre-internet communication, the body must be co-present or must be represented in fragments which are taken to identify a subjectivity with that body. In internet-dependent communication, the generation of a characteristic way of interacting that chimes with content and who a subjectivity claims they are is the way of accepting a particular identity.

A metaphor of signatures between online and offline when viewed through the rise of digital signatures repeats the story told at the start of this book. At first it might seem obvious that a digital signature, authorized by encryption, will be able to repeat the function of a handwritten signature. But then as the nature of digital signatures is explored difficulties arise and it turns out that calling it a signature in the same sense as an identity marker would be misleading. Instead, the digital signature has to be buttressed either by a pre-internet identity marker in physical co-presence or by the characteristic internet practice of judging a style of interaction. The communicative practices of pre-internet and internet-dependent transmission of messages can be seen in operation within digital signatures.

Both communicative practices analysed in relation to, on the one hand, nineteenth-century letters and, on the other hand, persistent virtual-world gaming include both senses of signature. In both there are markers of identity and there are particular ways of doing things that create an identity. The practices outlined are complex and offer extensive resources and repertoires of action through which the everyday moments of transmission are lived; that is, through which someone takes part in communication by sending or receiving a message in a way that tells them who is sending or receiving, offers confidence that the message has remained the same from sending to receiving, that tells them where to start and stop looking for the meaning of the message and allows all this to be assumed, to be enacted in the blink of an eye, and to be repeated. It is not the fact of signatures of both these types being present in different communicative practices that is at stake, this should be taken for granted. Instead it is centrally an issue of the ways they function and the elements they function in relation to. Clearly, an email address may function as an identity marker, a signature in the sense of object, but within internet-dependent communicative practices this object will always be suspect and secondary to the signature that a characteristic style of writing gives to someone's emails. And the same point is true of pre-internet communicative practices in which it is not so much the fact that there are elements that appear similar to those in internet-dependent practices but the fact that the practices constitute sets of links and arrangements, assemblages even, that situate elements in relation to each other.

This emphasis on interrelations and their workings, which emphatically does not mean they are static, unchanging or unchangeable, underlines how deeply anxiety will be embedded within communication because similar elements will continue to be used and mistaken for each other. Communicative anxiety takes hold deeply within transmission as the elements that seem to be the same are generated, constrained and managed from different practices which differently generate these elements. Normalized within the growing widespread familiarity with living with two communicative practices will be precisely this anxiety in which subjectivities particularly, but also meaning and messages, appear to be the same but derive from different material performatives which contradict and dismantle each other.

Signatures close this argument about how to understand communication after the rise of the internet. We live now with signatures in both senses of identity marker and of characteristic style, within two communicative practices that are both necessary and which contradict each other. We are after the internet, not in the sense of being without the internet or having seen the internet superseded, but because we now assume the internet alongside pre-existing communicative practices. We can now communicate due to two sets of communicative practices that each produce a presence from which we are able to create transmission.

Notes

1 Sir William R. Campion, KCMG, DSO, Governor of Western Australia, 1924–31 (Campion, 1939, p. xiii).

2 Fifty years of the internet depends on where it is counted as starting. Arpanet, the immediate predecessor, was 1969, making 2019 fifty years, but the invention of the Internet Protocol was not till 1983 making 2033 fifty years. No doubt other founding moments could be imagined.

Bibliography

Primary Sources

State Library of Victoria, Melbourne Australia

Meyrick: MS 7959 H15789–H15816

Black: MS8996 Box 3 Niel Black Colonial Correspondence Inwards 1850s

— MS8996 Box 20 Niel Black Inward Correspondence, Letters from Abroad 1840s

— MS8996 Box 25 Niel Black Outward Correspondence

Port Phillip Association: MS 1433–47 Port Phillip Association

Miscellaneous: MS Box 110/4b, MS 13530, MS56676, MS56699, MS56678, MS56681, MS56680, MS 6676–MS 6684, MS6682, MS9825, MS 13166

Mitchell Library, Sydney Australia

M 835 (microfilm), MLMSS 5518 (microfilm), A 290 (Safe 1/304), A 291 (Safe 1/305), A 292 (Safe 1/306), A 293 (Safe 1/307), A 294 (Safe 1/308)

Secondary Sources

Abu-Lughod, L. (1988), 'Fieldwork of a dutiful daughter', in S. Altorki and C. Fawzi El-Solh (eds), *Arab Women in the Field: Studying Your Own Society*. Syracuse, NY: Syracuse University Press, pp. 45–72.

Abunimah, A. (2011), 'New evidence about Amina, the "gay girl in damascus" hoax', *The Electronic Intifada Blog*. 12 June 2011, available at http://electronicintifada.net/blog/ali-abunimah/new-evidence-about-amina-gay-girl-damascus-hoax, accessed June 2011.

Addley, E. (2011), 'Syrian lesbian blogger revealed conclusively to be a married man'. *Guardian Newspaper*, 13 June 2011, available at http://www.guardian.co.uk/world/2011/jun/13/syrian-lesbian-blogger-tom-macmaster, accessed June 2011.

—(2010), 'MasterCard site partially frozen by hackers in WikiLeaks "revenge."' *Guardian Newspaper*. 8 December 2010, available at http://www.guardian.co.uk/media/2010/dec/08/mastercard-hackers-wikileaks-revenge, accessed August 2011.

Agamben, G. (2004), *The Open: Man and Animal*. Stanford: Stanford University Press.

—(1999), *Remnants of Auschwitz: The Witness and the Archive*. New York: Zone Books.

Altorki, S. and Fawzi El-Solh, C. (eds) (2006), *Arab Women in the Field: Studying Your Own Society*. Syracuse, NY: Syracuse University Press.

Araf, A. (2011), 'Ana Min Virginiya', *A Gay Girl in Damascus Blog*. 5 June 2011, accessed via Google Cache at http://webcache.googleusercontent.com/search?q=cache:kV22cNQPiHQJ:damascusgaygirl.blogspot.com/2011/06/ana-min-virginiya.html+I+am+complex,+I+am+many+things%3B+I+am+an+Arab,+I+am+Syrian,+I+am+a+woman,+I+am+queer,&cd=2&hl=en&ct=clnk&gl=uk&client=firefox-a&source=www.google.co.uk, in June 2011.

Barad, K. (2007), *Meeting the Universe Halfway: Quantum Physics and the Entanglement of Matter and Meaning*. London: Duke University Press.

Barnes, B., Bloor, D. and Henry, J. (1996), *Scientific Knowledge: A Sociological Analysis*. Chicago: Chicago University Press.

Baym, N. (2010), *Personal Connections in the Digital Age*. Cambridge: Polity.

Benedikt, M. (1992), 'Introduction', in M. Benedikt (ed.), *Cyberspace: First Steps*. Cambridge: MIT Press, pp. 1–25.

Benedikt, M. (ed.) (1992), *Cyberspace: First Steps*. Cambridge: MIT Press.

Bennett, T. and Watson, D. (eds) (2002), *Understanding Everyday Life*. Oxford: Blackwell.

Bernasconi, R. and Wood, D. (eds) (1988), *The Provocation of Levinas: Rethinking the Other*. London: Routledge.

Bloor, D. (1997), *Wittgenstein, Rules and Institutions*. London: Routledge.

—(1983), *Wittgenstein: A Social Theory of Knowledge*. London: MacMillan.

Butler, J. (2005), *Giving an Account of Oneself*. New York: Fordham University Press.

—(1997), *Excitable Speech: A Politics of the Performative*. London: Routledge.

—(1993), *Bodies That Matter: On the Discursive Limits of 'Sex'*. London: Routledge.

Cadava, E., Connor, P. and Nancy, J.-L. (eds) (1991), *Who Comes after the Subject?* London: Routledge.

CAE (1996), *Electronic Civil Disobedience and Other Unpopular Ideas*. New York: Autonomedia.

Callon, M. and Latour, B. (1981), 'Unscrewing the big leviathan: how actors macro-structure reality and how sociologists help them to do so', in K. Knorr-Cetina and A. Cicourel (eds), *Advances in Social Theory and Methodology: Towards An Integration of Micro- and Macro-Sociologies*. London: Routledge and Kegan Paul, pp. 277–303.

Campion, W. (1939), 'Foreward', in F. J. Meyrick (ed.), *Life in the Bush (1840–1847): A Memoir of Henry Howard Meyrick*. Sydney: T.Nelson and Sons, pp. xiii–xvi.

Carey, J. W. (2009), *Communication as Culture* (revised edition). London: Routledge.

Caron, A. and Caronia, L. (2007), *Moving Cultures: Mobile Communication in Everyday Life*. London: McGill-Queen's University Press.

Castronova, E. (2007), *Exodus to the Virtual World: How Online Fun Is Changing Reality*. Basingstoke: Palgrave Macmillan.

—(2006), *Synthetic Worlds: The Business and Culture of Online Games*. Chicago: University of Chicago Press.

Chanter, T. (1988), 'Feminism and the Other', in R. Bernasconi and D. Wood (eds), *The Provocation of Levinas: Rethinking the Other*. London: Routledge, pp. 32–56.

Cherny, L. and Weise, E. (eds) (1996), *Wired_Women: Gender and New Realities in Cyberspace*. Seattle: Seal Press.

Clanchy, M. T. (1993), *From Memory to Written Record: England 1066–1307* (second edition). Oxford: Blackwell.

Collins, H. M. (1992), *Changing Order: Replication and Induction in Scientific Practice*. Chicago: Chicago University Press.

Corneliussen, H. and Rettberg, J. W. (eds) (2008), *Digital Culture, Play, and Identity: A World of Warcraft Reader*. Cambridge: MIT Press.

Craig, R. T. (2007), 'Pragmatism in the field of communication theory.' *Communication Theory*,17(2), 125–45.

—(2006), 'Communication as a practice', in G. J. Shepherd, St. J. John and T. Striphas (eds), *Communication as . . .: Perspectives on Theory*. Thousand Oaks, CA: Sage, pp. 38–47.

—(1999), 'Communication theory as a field'. *Communication Theory*, 9(2), 119–61.

Crossley, N. (2006), *Reflexive Embodiment in Contemporary Society*. Maidenhead: Open University.

de Certau, M. (1984), *The Practice of Everyday Life*. Berkeley: University of California Press.

Deetz, S. A. (1994), 'Future of the discipline', in S. A. Deez (ed.), *Communication Yearbook 17*. London: Sage, pp. 565–600.

Deez, S. A. (ed.) (1994), *Communication Yearbook 17*. London: Sage.

Deleuze, G. (1994), *Difference and Repetition*. London: Athlone Press.

Dent, G. (2010), 'The things I've learned about teenagers'. *Guardian Newspaper*; G2, 30 August 2010, 10–11.

Derrida, J. (1991), '"Eating well" or the calculation of the subject: an interview with Jacques Derrida', in E. Cadava, P. Connor and J.-L. Nancy (eds), *Who Comes after the Subject?* London: Routledge, pp. 96–119.

—(1989), *On Spirit: Heidegger and the Question*. Chicago: Chicago University Press.

—(1988), *Limited Inc*. Evanston: Northwestern University Press.

—(1982a), 'Signature event context', in J. Derrida (ed.), *Margins of Philosophy*. Chicago: Chicago University Press, pp. 307–30.

—(1982b), 'Différance', in J. Derrida (ed.), *Margins of Philosophy*. Chicago: University of Chicago Press, pp. 1–28.

—(1982c), *Margins of Philosophy*. Chicago: Chicago University Press.

—(1981), *Positions*. London: Athlone Press.

—(1976), *Of Grammatology*. Baltimore: Johns Hopkins University Press.

—(1973), *Speech and Phenomena and Other Essays on Husserl's Theory of Signs*. Evanston: Northwestern University Press.

Dery, M. (ed.) (1994), *Flamewars: The Discourse of Cyberculture*. Raleigh: Duke University Press.

Donath, J. (1998), 'Identity and deception in the virtual community', in P. Kollock and M. Smith (eds), *Communities in Cyberspace*. London: Routledge, pp. 29–59.

Dreyfus, H. (1995), *Being-in-the-World: A Commentary on Heidegger's Being and Time, Division 1*. Cambridge: MIT Press.

Dreyfus, H. and Wrathall, M. (eds) (2005), *A Companion to Heidegger: Blackwell Companions to Philosophy*. Oxford: Blackwell.

du Gay, P., Evans, J. and Redman, P. (eds) (2000), *Identity: A Reader*. London: Sage.

Featherstone, M. and Burrows, R. (1995), 'Cultures of technological embodiment: an introduction', in M. Featherstone and R. Burrows (eds), *Cyberspace/Cyberbodies/Cyberpunk*. London: Sage, pp. 1–20.

Gautney, H. (2010), *Protest and Organization in the Alternative Globalization Era: NGOs, Social Movements, and Political Parties*. Basingstoke: Palgrave Macmillan.

Goldstein, E. (ed.) (2008), *The Best of 2600: A Hacker Odyssey*. Indianapolis: Wiley.

Goodin, D. (2010), 'Virtual sit-in tests line between DDOS and free speech', *The Register*. 9 April 2010, available at http://www.theregister.co.uk/2010/04/09/virtual_protest_as_ddos/, accessed January 2012.

Gregory, J. (ed.) (2005), *A Companion to Greek Tragedy*. Oxford: Blackwell.

Guardian-MEL (2011), 'Middle East Live', *Guardian Newspaper Blog*. 8 June 2011, available at http://www.guardian.co.uk/world/middle-east-live/2011/jun/08/syria-middle-east-unrest, accessed June 2011.

Hagström, C. (2008), 'Gaming and naming in World of Warcraft', in H. Corneliussen and J. W. Rettberg (eds), *Digital Culture, Play, and Identity: A World of Warcraft Reader*. Cambridge: MIT Press, pp. 266–85.

Haraway, D. (2008), *When Species Meet*. London: University of Minnesota Press.

—(1991), *Simians, Cyborgs, and Women: The Reinvention of Nature*. London: Free Association Books.

Heidegger, M. (1980), *Being and Time*. Oxford: Basil Blackwell.

Ihde, D. (2002), *Bodies in Technology*. London: University of Minnesota Press.

Joinson, A. (2003), *Understanding the Psychology of Internet Behaviour: Virtual Worlds, Real Lives*. Basingstoke: Palgrave Macmillan.

Jonas, H. (1964), 'Heidegger and theology'. *The Review of Metaphysics*, 18(2), 207–33.

Jones, S. (1995), 'Introduction: from where to who knows?', in S. Jones (ed.), *Cybersociety: Computer-Mediated Communication and Community*. London: Sage, pp. 1–9.

Jordan, T. (2008), *Hacking: Digital Media and Technological Determinism*. Cambridge: Polity Press.

—(1999), *Cyberpower: The Culture and Politics of Cyberspace and the Internet*. London: Routledge.

—(1995), 'The unity of social movements'. *Sociological Review*, 23(4), 125–44.

Jordan, T. and Taylor, P. (2004), *Hacktivism and Cyberwars: Rebels with a Cause?* London: Routledge.

—(1998), 'A sociology of hackers'. *Sociological Review*, 46(4), 757–80.

Kaminski, M. (2004), *Games Prisoners Play*. Princeton: Princeton University Press.

Kiddle, M. (1961), *Men of Yesterday*. Melbourne: Melbourne University Press.

King, B. and Borland, J. (2003), *Dungeons and Dreamers: The Rise of Computer Game Culture, from Geek to Chic*. Emeryville: McGraw Hill.

Kinross-Smith, G. (2002), 'Neil Black of Glenormiston', *Redreaming the Future: Online Resource*, available at http://www.redreaming.info/DisplayStory.asp?id=89, accessed October 2010.

Knorr-Cetina, K. and Cicourel, A. (eds) (1981), *Advances in Social Theory and Methodology: Towards An Integration of Micro- and Macro-Sociologies*. London: Routledge and Kegan Paul.

Kollock, P. and Smith, M. (eds) (1998), *Communities in Cyberspace*. London: Routledge.

Kovacs, D. (2005), 'Chapter twenty-four text and transmission', in J. Gregory (ed.), *A Companion to Greek Tragedy*. Oxford: Blackwell, pp. 379–92.

Lacan, J. (1977a), *Écrits: A Selection*. London: Tavistock Publications.

—(1977b), *The Four Fundamental Concepts of Psycho-Analysis*. Harmondsworth: Penguin.

Lash, S. (2002), *Critique of Information*. London: Sage.

Latour, B. (2005a), 'From realpolitick to dingpolitik', in B. Latour and P. Wiebel (eds), *Making Things Public – Atmospheres of Democracy*. Cambridge: MIT Press.

—(2005b), *Reassembling the Social: An Introduction to Actor-Network-Theory*. Oxford: Oxford University Press.

—(1992), 'Where are the missing masses: sociology of a door', available at http://www.bruno-latour.fr/articles/article/050.html, accessed February 2010.

Latour, B. and Wiebel, P. (eds) (2005), *Making Things Public – Atmospheres of Democracy*. Cambridge: MIT Press.

Latour, B. and Woolgar, S. (1986), *Laboratory Life: The Construction of Scientific Facts*. Princeton: Princeton University Press.

Lee, A. (2011), '4Chan Founder: Mark Zuckerberg is "totally wrong" about anonymity.' *The Huffington Post*. 15 March 2011, available at http://www.huffingtonpost.com/2011/03/14/4chan-founder-mark-zuckerberg_n_835424.html, accessed June 2011.

Levinas, E. (2006), *Humanism of the Other*. Urbana: University of Illinois Press.

—(1981), *Otherwise Than Being or Beyond Essence*. Pittsburgh: Duquesne University Press.

—(1969), *Totality and Infinity: An Essay on Exteriority*. Pittsburgh: Duquesne University Press.

Levy, P. (1998), *Becoming Virtual: Reality in the Digital Age*. London: Plenum Trade.

Levy, S. (2000), *Crypto: Secrecy and Privacy in the New Code War*. London: Penguin.

Licoppe, C. (2005), '"Connected Presence": the emergence of a new repertoire for managing social relationships in a changing communication technoscape'. *Environment and Planning D: Society and Space*, 22, 135–56.

Ling, R. and Donner, J. (2009), *Mobile Communication*. Cambridge: Polity.

Loader, B. (1997), 'The Governance of Cyberspace: Politics, Technology and Global Restructuring', in B. Loader (ed.), *The Governance of Cyberspace: Politics, Technology and Global Restructuring*. London: Routledge, pp. 1–22.

Loxley, J. (2007), *Performativity*. London: Routledge.

Luhman, N. (1995), *Social Systems*. Stanford: Stanford University Press.

MacCullam-Stewart, E. and Parsler, J. (2008), 'Role-play versus gameplay: the difficulties of playing a role in World of Warcraft', in H. Corneliussen and J. W. Rettberg (eds), *Digital Culture, Play, and Identity: A World of Warcraft Reader*. Cambridge: MIT Press, pp. 225–45.

Macintyre, S. (2004), *A Concise History of Australia*. Cambridge: Cambridge University Press.

MacKellar, M. (2008), *Strangers in a Foreign Land: The Journal of Niel Black and Other Voices from the Western District*. Melbourne: Miegunyah Press.

Malaby, T. (2009), *Making Virtual Worlds: Linden Lab and Second Life*. Ithaca: Cornell University Press.

Markoff, J. (2005), *What the Doormous Said: How the Sixties Counter-Culture Shaped the Personal Computer Industry*. Harmondsworth: Penguin.

McCarthy, C., Packer, J. and Robertson, C. (eds) (2005), *Thinking with James Carey: Essays on Communications, Transport, History*. New York: Peter Lang Publishing.

Melucci, A. (1996), *The Playing Self: Person and Meaning in the Planetary Society*. Cambridge: Cambridge University Press.

Menn, J. (2010), *Fatal System Error: The Hunt for the New Crime Lords Who Are Bringing Down the Internet*. New York: Public Affairs.

Meyrick, F. J. (1939), *Life in the Bush (1840–1847): A Memoir of Henry Howard Meyrick*. Sydney: T.Nelson and Sons.

Milne, E. (2010), *Letters, Postcards, Email: Technologies of Presence*. London: Routledge.

MW (2010), *Measuring Worth: Conversion of British Pounds from 1264 to Present*, available at http://www.measuringworth.com/ppoweruk/, accessed October 2010.

Newman, J. (2002), 'The myth of the egodic videogame'. *Game Studies*, 2(1), http://www.gamestudies.org/0102/newman/, accessed April 2011.

Ott, H. (1993), *Martin Heidegger: A Political Life*. London: HarperCollins.

Peters, J. D. (1999), *Speaking into the Air: A History of the Idea of Communication*. Chicago: Chicago University Press.

—(1993), 'Genealogical notes on "the field."' *Journal of Communication*, 43(4), 132–9.

Plato (1892), *The Dialogues of Plato, vol.1* (3rd edition revised). Oxford: Oxford University Press, available at http://oll.libertyfund.org/title/111, accessed July 2011.

Polt, R. (1999), *Heidegger: An Introduction*. London: Routledge.

Poulsen, K. (2011), *Kingpin: How One Hacker Took Over the Billion-Dollar Cybercrime Underground*. New York: Crown Publishers.

Redman, P. (2000), 'The subject of language, ideology and discourse: introduction,' in P. du Gay, J. Evans and P. Redman (eds), *Identity: A Reader*. London: Sage, pp. 9–14.

Rheingold, H. (1991), *Virtual Reality*. New York: Touchstone.

Rockmore, T. (1995), *Heidegger and French Philosophy: Humanism, Antihumanism and Being*. London: Routledge.

Rowling, J. K. (1998), *Harry Potter and the Chamber of Secrets*. London: Bloomsbury.

Schatzki, T. (2005), 'Early Heidegger and sociality,' in H. Dreyfus and M. Wrathall (eds), *A Companion to Heidegger: Blackwell Companions to Philosophy*. Oxford: Blackwell, pp. 233–47.

Seale, C. (ed.) (1999), *Researching Society and Culture*. London: Sage.

Serres, M. (2008), *The Five Senses: A Philosophy of Mingled Bodies (I)*. London: Continuum Books.

Shannon, C. E. (1948), 'A mathematical theory of communication'. *The Bell Systems Technical Journal*, 27, 379–424 and 623–56.

Shepherd, G. J., St. John, J. and Striphas, T. (eds) (2006), *Communication as . . .: Perspectives on Theory*. Thousand Oaks, CA: Sage.

Standage, T. (2007), *The Victorian Internet: The Remarkable Story of the Telegraph and the Nineteenth Century's On-Line Pioneers*. New York: Walker and Company.

Staten, H. (1986), *Wittgenstein and Derrida*. Lincoln: University of Nebraska Press.

Stengers, I. (2005), 'Whitehead's account of the sixth day'. *Configurations*, 13, 35–55.

Sterling, B. (1994), *The Hacker Crackdown: Law and Disorder on the Electronic Frontier*. London: Penguin.

Stone, A. R. (1995), *The War of Desire and Technology at the Close of the Mechanical Age*. Cambridge: MIT Press.

Taylor, P. (1999), *Hackers: Crime in the Digital Sublime*. London: Routledge.

Thomson, I. (2005), 'Heidegger and national socialism', in H. Dreyfus and M. Wrathall (eds), *A Companion to Heidegger: Blackwell Companions to Philosophy*. Oxford: Blackwell, pp. 32–48.

Touraine, A. (1995), *Critique of Modernity*. Oxford: Blackwell.

Tronstad, R. (2008), 'Character identification in World of Warcraft: the relationship between capacity and appearance', in H. Corneliussen and J. W. Rettberg (eds), *Digital Culture, Play, and Identity: A World of Warcraft Reader*. Cambridge: MIT Press, pp. 249–63.

Turkle, S. (1996), *Life on the Screen: Identity in the Age of the Internet*. London: Phoenix.

Turner, F. (2006), *From Counterculture to Cyberculture: Stewart Brand, the Whole Earth Network and the Rise of Digital Utopianism*. Chicago: Chicago University Press.

Vlastos, G. (1994), *Socratic Studies*. Cambridge: Cambridge University Press.

Walsh, D. (1999), 'Doing ethnography', in C. Seale (ed.), *Researching Society and Culture*. London: Sage, pp. 217–32.

Ward, R. (2010), 'Black, Niel (1804–1880)', *Australian Dictionary of Biography: Online Edition*, available at http://adbonline.anu.edu.au/biogs/A030163b.htm, accessed October 2010.

Weber, S. (2004), *The Success of Open Source*. Harvard: Harvard University Press.

Winterhalter, R. (2011), 'Nintendo hacked but personal data safe'. *1Up.com*, available at http://www.1up.com/news/nintendo-hacked-personal-data-safe, accessed February 2012.

Wittgenstein, L. (1958), *Philosophical Investigations*. Oxford: Basil Blackwell.

Index